Collins

Weekly Reasoning Tests

Year 6/2nd Level for P7/S1

Samantha Townsend

William Collins' dream of knowledge for all began with the publication of his first book in 1819. A self-educated mill worker, he not only enriched millions of lives, but also founded a flourishing publishing house. Today, staying true to this spirit, Collins books are packed with inspiration,
innovation and practical expertise. They place you at the centre of a world of possibility and give you exactly what you need to explore it.

Collins. Freedom to teach.

Collins
An imprint of HarperCollins*Publishers*
The News Building
1 London Bridge Street
London
SE1 9GF

Browse the complete Collins catalogue at
www.collins.co.uk

MIX
Paper from responsible sources
FSC™ C007454

This book is produced from independently certified FSC™ paper to ensure responsible forest management.

For more information visit: www.harpercollins.co.uk/green

© HarperCollinsPublishers Limited 2019

10 9 8 7 6 5 4 3 2 1

ISBN 978-0-00-833341-6

All rights reserved. No part of this publication may be reproduced, stored in a retrieval system,
or transmitted in any form by any means, electronic, mechanical, photocopying, recording or otherwise, without the prior written permission of the Publisher or a licence permitting restricted copying in the United Kingdom issued by the Copyright Licensing Agency Ltd., Barnard's Inn, 86 Fetter Lane, London, EC4A 1EN.

British Library Cataloguing in Publication Data. A catalogue record for this publication is available from the British Library.

Author: Samantha Townsend
Reviewer: Paul Broadbent
Publisher: Katie Sergeant
Commissioning Editor: Fiona Lazenby
Product Developer: Mike Appleton
Copyeditor: Gwynneth Drabble
Proofreader: Catherine Dakin
Design and Typesetting: Ken Vail Graphic Design
Cover Design: The Big Mountain Design
Production controller: Katharine Willard

Contents

Introduction iv

Reasoning Test 1   1
Reasoning Test 2   7
Reasoning Test 3   14
Reasoning Test 4   22
Reasoning Test 5   28
Reasoning Test 6   35
Reasoning Test 7   42
Reasoning Test 8   48
Reasoning Test 9   54
Reasoning Test 10   60
Reasoning Test 11   65
Reasoning Test 12   71
Reasoning Test 13   76
Reasoning Test 14   82
Reasoning Test 15   90
Reasoning Test 16   96
Reasoning Test 17   103
Reasoning Test 18   110
Reasoning Test 19   115
Reasoning Test 20   121
Reasoning Test 21   129
Reasoning Test 22   135
Reasoning Test 23   141
Reasoning Test 24   149
Reasoning Test 25   154
Reasoning Test 26   161
Reasoning Test 27   168
Reasoning Test 28   175
Reasoning Test 29   183
Reasoning Test 30   192

Answers 200

Record Sheet 235

Introduction

Collins Assessment Weekly Reasoning Tests provide 30 weekly photocopiable tests for Year 6/P7 to help you uncover gaps in mathematical reasoning and problem-solving skills. The tests are designed to reflect the style and type of questions from the English National Curriculum tests (SATs) for Key Stage 2 Mathematics Paper 2 and Paper 3: Reasoning to help children become familiar and confident with the format. They will also help children in Scotland prepare for the P7 numeracy Scottish National Standardised Assessment.

How to use this book

Each test has **12 questions** worth **18 marks** in total, including a variety of 1-mark and 2-mark questions with a 3-mark question as the final question in each test. Each test starts with an accessible 1-mark question and becomes progressively more challenging as pupils work through the questions. The tests should take **20 minutes** to complete to help pupils practise their time management in test situations.

Each test presents mathematical problems in a range of formats to ensure pupils can demonstrate mathematical fluency, problem solving and reasoning as well as experience the most common types of problem-solving questions: multi-step word problems, missing number problems, finding all possibilities, logic puzzles, visualisation puzzles and reasoning problems (for example, explaining why a statement is true/false). They include both selected response questions (e.g. multiple choice, matching, yes/no) and constructed response questions. Questions draw on all content domains and approximately half of the questions in each test are presented in context.

Each test follows a similar structure with the different areas of focus appearing in the same position within each test. This will help pupils see an improvement over time and enable teachers to quickly identify problem-solving and reasoning skills with which pupils may require support and further input when approaching these question types. Questions draw on different content domains from test to test, so pupils can apply their reasoning skills to a wide variety of mathematical topics across the tests.

Test structure

Question	Area of focus
1	Known facts/Mathematical terms
2	Data interpretation
3	Measures construction
4	Geometry
5	Fractions calculation
6	Fractions equivalence
7	Explain how you know
8	Show your method multi-step problem
9	Algebraic reasoning
10	Show your method multi-step problem
11	Percentages/Ratio/Proportion/Scale factors
12	Show your method multi-step problem

Answers

The answers for the tests are provided in the mark schemes at the back of the book. For each question the mark scheme includes the content domain reference, answer requirement, marks to be awarded, plus additional guidance on what to accept/not accept and when to award method marks in multi-mark questions where pupils do not achieve the correct final answer.

Recording progress

Documents to help you record progress are provided online with the free download that accompanies this book. You can use the pupil record sheet (also included at the back of the book) to provide evidence of which areas each child has performed well in and where he/she needs to focus to develop their reasoning skills. You can easily record results for your classes on the spreadsheet and identify any common gaps in understanding. The spreadsheet can then be used to inform your next teaching and learning steps.

Editable download

All the tests are available online in Word and PDF format. Go to collins.co.uk/assessment/downloads to find instructions on how to download. The files are password protected and the password clue is included on the website. You will need to use the clue to locate the password in your book.

Reasoning Test 1

Name _____

1. In this sequence, the numbers increase by the same amount each time.
Write the missing numbers in the boxes.

  64

1 mark

2. This bar chart shows the different ways in which children travel to school.

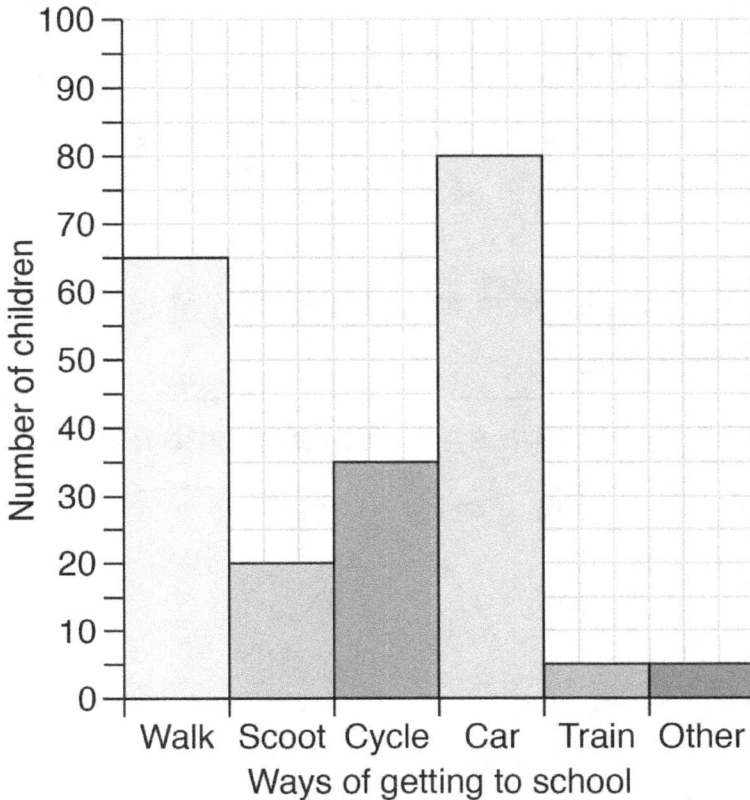

a) How many children responded?

b) How many children did not go to school by car?

1 mark

Reasoning Test 1

Name _____

3 Mary goes to work. These clocks show when she leaves home in the morning and when she returns home in the evening.

Time Mary leaves home  Time Mary returns home

How long is Mary out of the house?

| hours | minutes |

1 mark

4 Tick the shapes that would go in the intersection of the Venn diagram. One has been done for you.

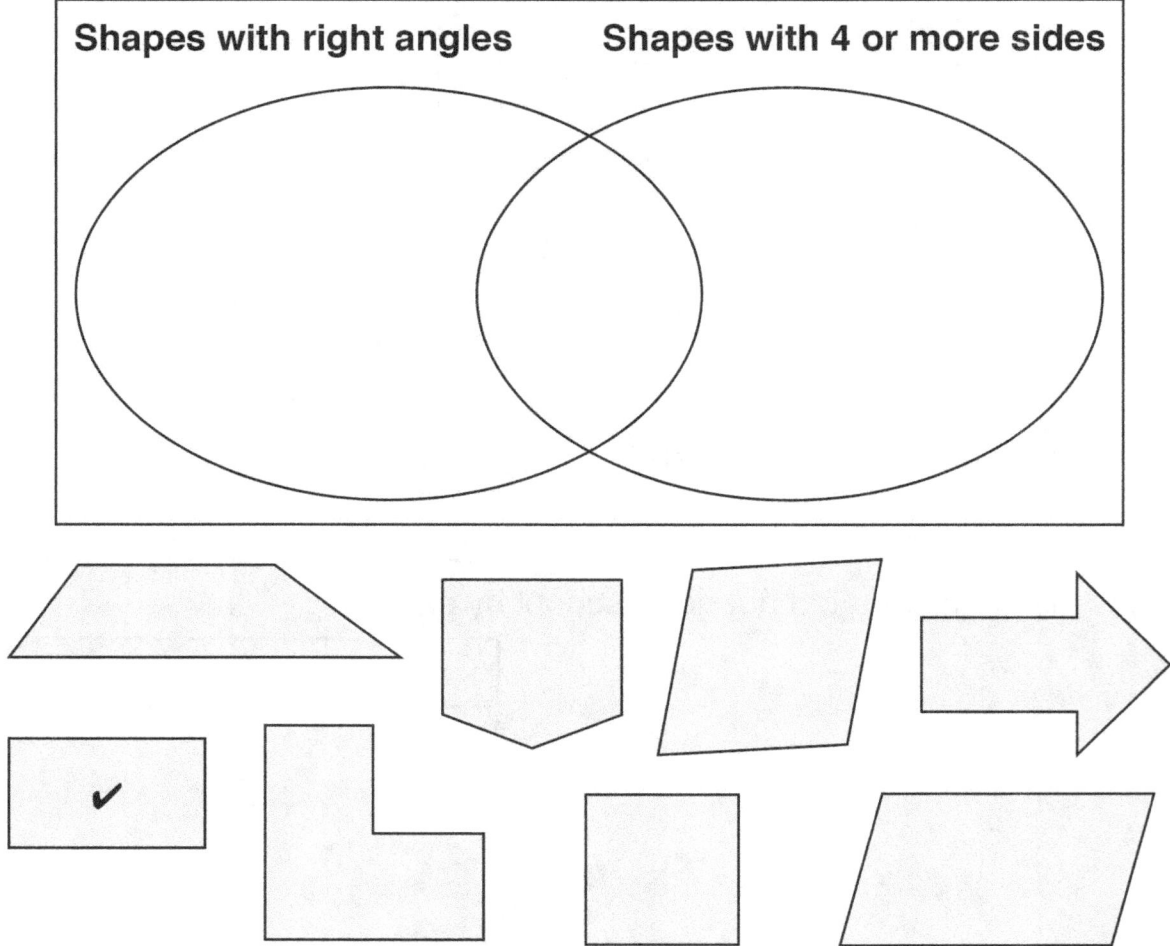

1 mark

Reasoning Test 1   Name _____

5 Here are four fraction cards.

$\boxed{\dfrac{3}{4}} \quad \boxed{\dfrac{5}{8}} \quad \boxed{\dfrac{7}{14}} \quad \boxed{\dfrac{4}{16}}$

Use any three of the cards to make this correct.

☐ < ☐ < ☐

1 mark

6 Order the fractions below, from smallest to largest.

$\dfrac{1}{2} \quad \dfrac{7}{8} \quad \dfrac{4}{2} \quad \dfrac{2}{8} \quad \dfrac{3}{4}$

☐ ☐ ☐ ☐ ☐

smallest   **largest**

1 mark

7 Veejay is laying wooden flooring throughout his house.

He needs 2,054 pieces of wood for downstairs. He needs 1,880 pieces for upstairs.

Veejay has ordered 5,000 pieces.

His sister would like wooden flooring in her lounge. She needs 1,200 pieces of wood.

Will Veejay have enough wood left to cover his sister's lounge floor when he has finished his own house?

Explain your reasoning below.

1 mark

Reasoning Test 1   Name _____

8   Large pizzas cost £9.50 each.

Small pizzas cost £7.50 each.

Five children together buy two large pizzas and three small pizzas.

They share the cost equally.

How much does each child pay?

Show your method

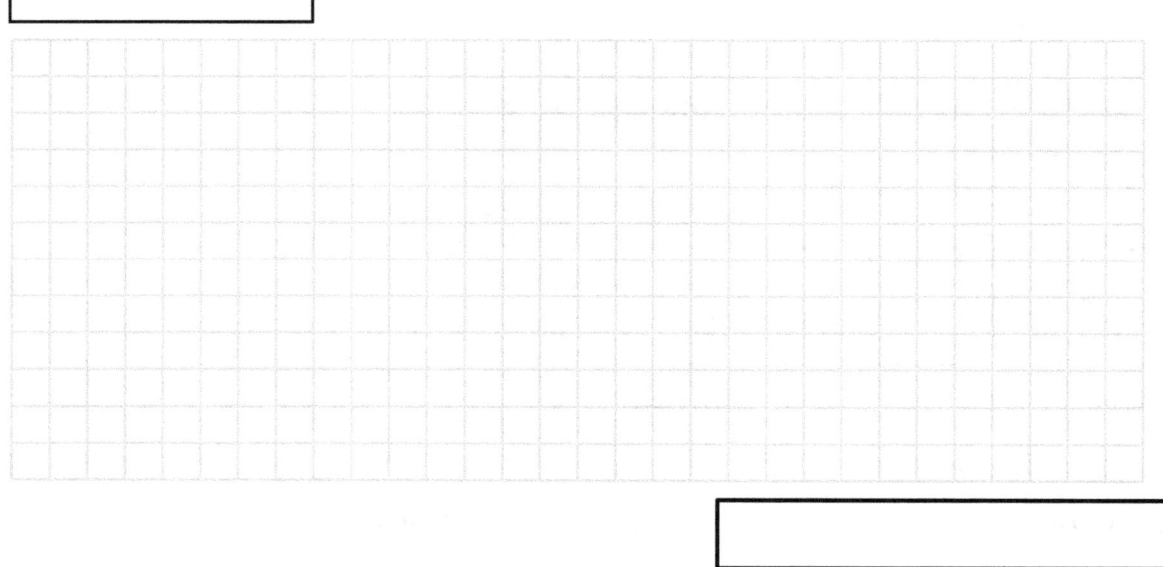

2 marks

9   A shaded isosceles triangle is drawn inside a rectangle.

What is the size of angle *a*?

Show your method

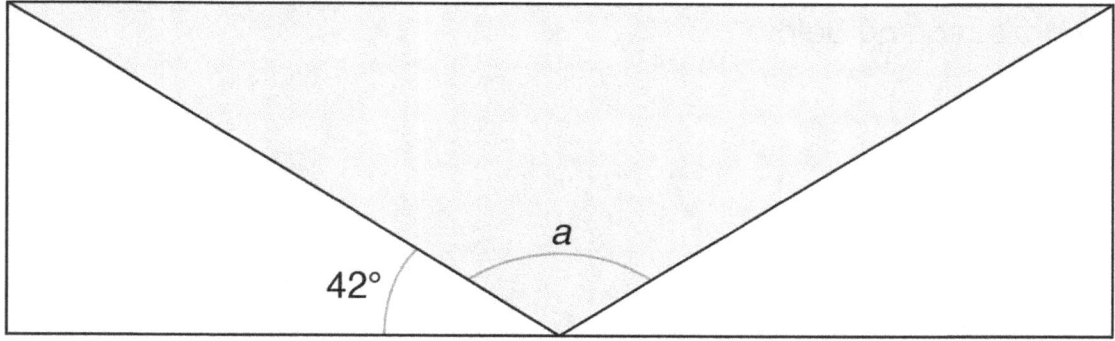

42°

Not drawn accurately

2 marks

10 Maya and her brother are saving for a new television.

Maya has saved £120. Her brother has saved 10% more than she has.

A television that usually costs £350 has 30% off its price, in the sale.

Do Maya and her brother have enough money to buy the television?

2 marks

11 A chef makes 2.75 L of cheese sauce for a dish. For every 250 ml of sauce, he needs 40 g of cheese.

a) How much cheese does he need?

Reasoning Test 1   Name _____

When the chef finishes cooking, he has 15% of the cheese sauce left.

b) How many millilitres are left?

ml

2 marks

12 The area of a school playing field is 6,875 m².

The school football pitch is 108 m long and 45 m wide.

How much larger is the playing field than the school football pitch?

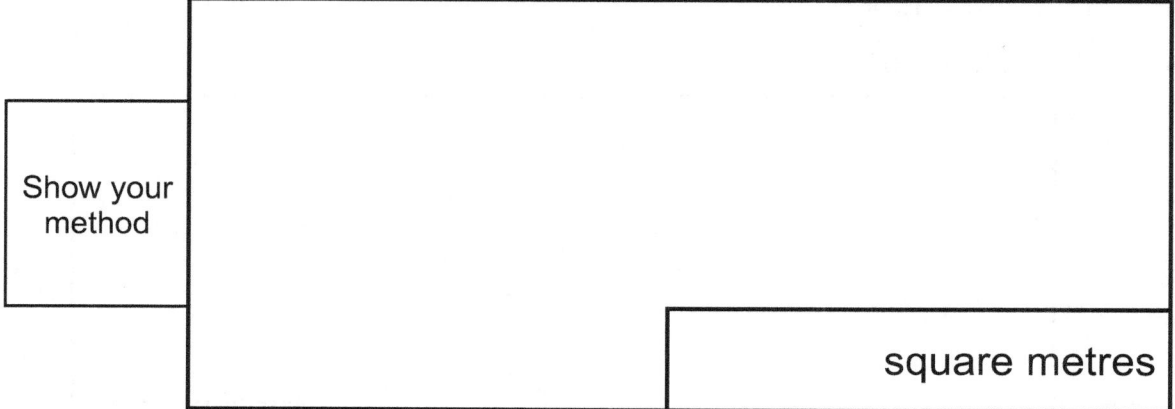

square metres

3 marks

Total marks ..../18

Reasoning Test 2

Name _____

1   Write the next three numbers in the sequence.

1 mark

2   This graph shows the average temperature in six cities during the month of March.

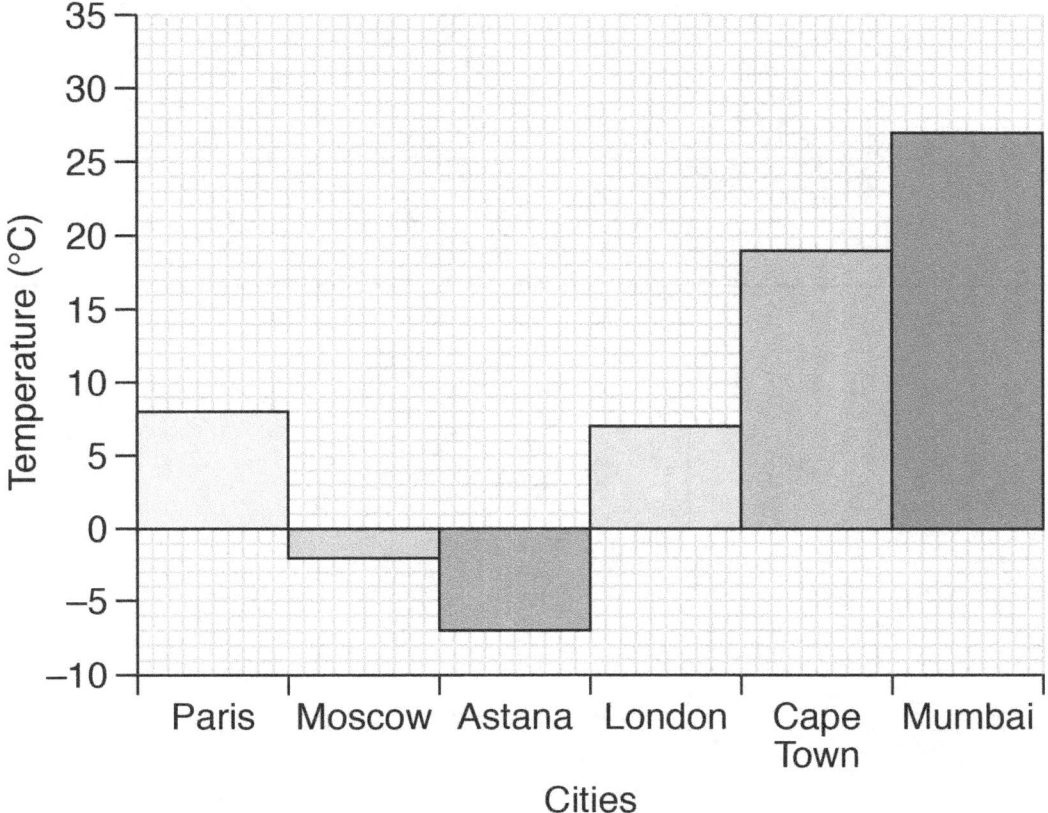

How many degrees higher is the temperature in Mumbai than in Astana in March?

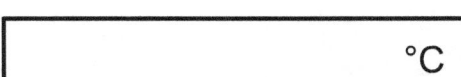 °C

1 mark

3 Calculate the perimeter of this shape.

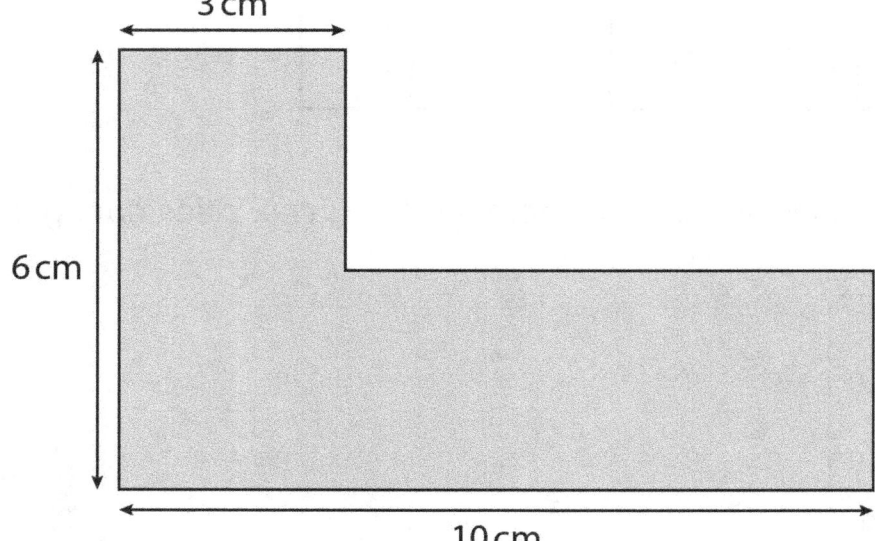

[ ] cm

1 mark

Reasoning Test 2   Name _____

4 Plot the coordinates on the grid and then join the points to create a polygon.

(1, 3)   (4, 3)   (5, 5)   (2, 5)

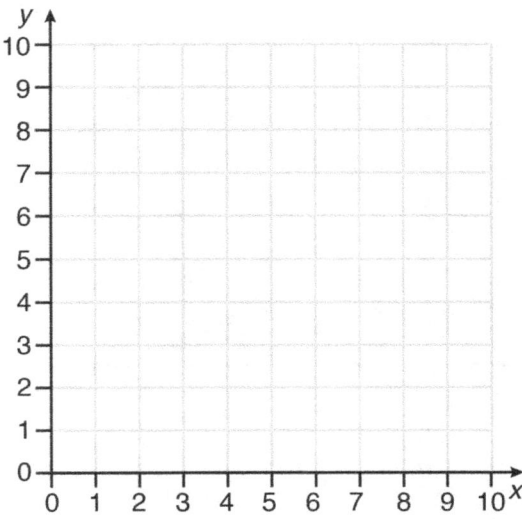

1 mark

5 Circle three fractions that are equivalent to $\frac{3}{4}$.

$\frac{5}{2}$ $\frac{4}{16}$ $\frac{3}{9}$ $\frac{5}{9}$ $\frac{21}{28}$ $\frac{75}{100}$ $\frac{2}{4}$ $\frac{18}{24}$

1 mark

6 Frankie spends £10 at the shop. The fraction next to each item shows how much of her £10 she used.

Write the amount of money Frankie spends on each item.

 $\frac{1}{5}$ £ ☐

 $\frac{3}{10}$ £ ☐

 $\frac{2}{5}$ £ ☐

 $\frac{1}{10}$ £ ☐

1 mark

Reasoning Test 2   Name _____

7   Jared says this is a trapezium, but Anna says it is not.
    Give three reasons to prove why this shape is not a trapezium.

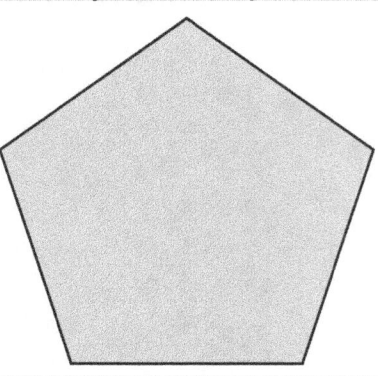

1 mark

8   Julia bakes scones for the school summer fayre. She works out how to price each box of scones like this:

    Cost = number of scones in box × 15p + 20p per box

    a) What should she price a box of 12 scones?

£ _____

Reasoning Test 2

Charlie buys a box of scones for £1.40.

b) Use Julia's pricing formula to calculate how many scones there are in Charlie's box.

2 marks

9 In which of these two equations is the value of b greater?

A $4b + 6 = 22$

B $3b - 7 = 29$

Explain your reasoning below.

2 marks

10 A flight from New York to London takes 7 hours.

a) If a plane departs from New York at 8 a.m. local time and arrives in London at 8 p.m. local time, how many hours ahead of New York is London?

Show your method

hours

Reasoning Test 2   Name _____

When it is refuelled and cleaned, the plane takes off at 10 p.m. from London for Dubai. It lands at 9 a.m. local time.

b) If Dubai is 4 hours ahead of London, how long does the flight take?

Show your method

_____ hours

2 marks

11 A decorator mixes green paint from pots of blue and yellow paint. For every 1 part blue, he uses 7 parts yellow.

The decorator is painting a room that requires 7 litres of green paint.

How many millilitres of blue and of yellow will he need to paint the room?

Show your method

blue _____ ml   yellow _____ ml

2 marks

Reasoning Test 2

12 Of the 96 children in Year 6, $\frac{3}{4}$ have pets. 45 children have dogs and 21 children have cats.

a) How many children have other types of pet but not a dog or a cat?

2 marks

b) What proportion of Year 6 children have other kinds of pet but not a cat or dog?

1 mark

Total marks ……../18

Reasoning Test 3   Name _____

1 Circle the square numbers.

56   4   80   77   25   14   49

1 mark

2 The table shows prices per night on a campsite in Cornwall.

Pitch type	Minimum size	Maximum occupancy (one family per pitch)	Low season	Mid season	High season
Hiking	6 m × 6 m	2	£12.00	£18.00	£24.00
Standard	10 m × 8 m	6	£18.00	£22.00	£36.00
Luxury	10 m × 8 m	6	£24.00	£33.00	£46.00
Deluxe	14 m × 10 m	6	£39.00	£48.00	£59.00
Essential	10 m × 8 m	6	£20.00	£25.00	£30.00
Dogs	Maximum 1		£4.00	£5.00	£6.00
Extra car	Maximum 1 extra		£4.00	£5.00	£6.00

How much would it cost for a family to stay on an essential pitch with their dog for 7 nights in low season?

£ _____

1 mark

3 Calculate the area of this triangle.

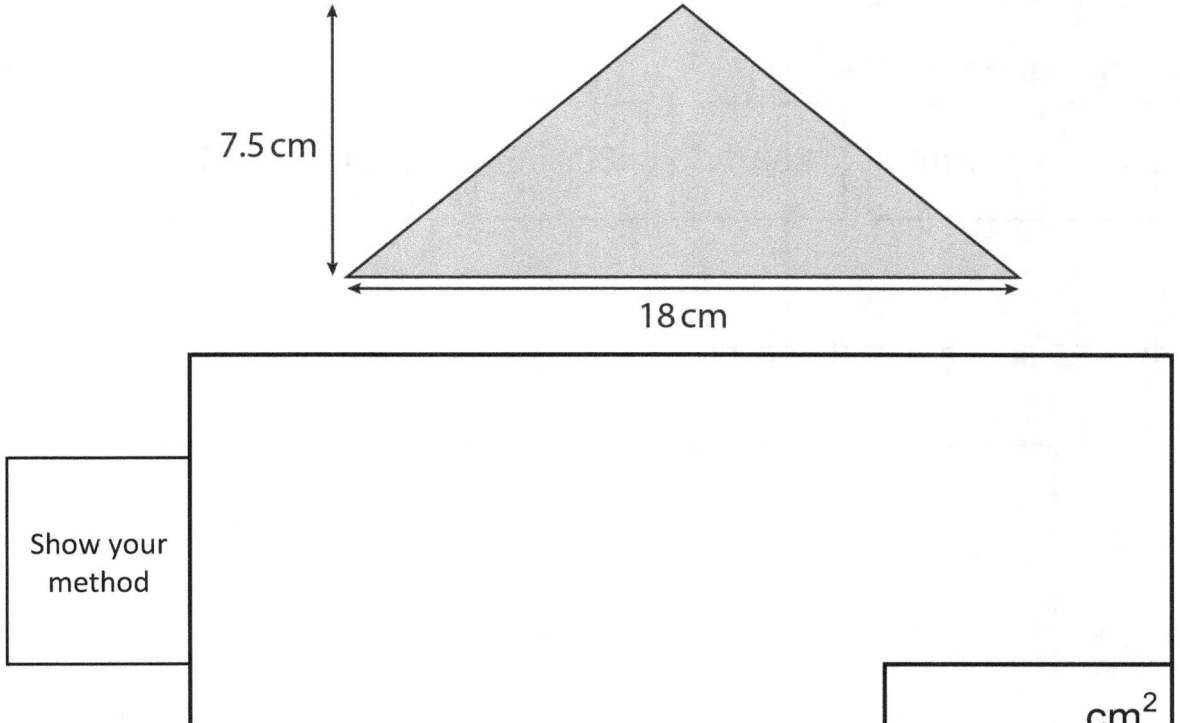

Show your method

cm²

1 mark

4 Circle the shape with a perimeter greater than 20 cm.

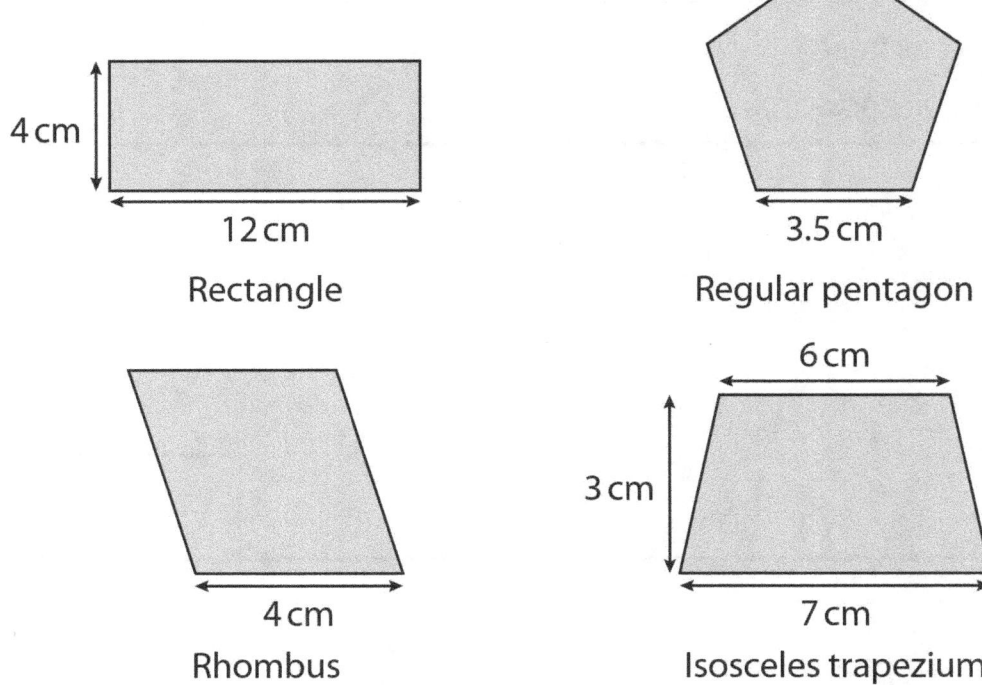

Rectangle

Regular pentagon

Rhombus

Isosceles trapezium

1 mark

Reasoning Test 3   Name _____

5 What is $\frac{2}{5}$ of two litres?

Circle the correct answer.

| 400 ml | 80 ml | 200 ml | 800 ml | 600 ml | 0.08 L |

1 mark

6 Express $\frac{3}{8}$ as a decimal fraction.

Show your method

1 mark

7 'The sum of three even numbers will never make an odd number.'

Is this true or false?

Explain your answer.

True / False

1 mark

Reasoning Test 3   Name _____

8 A box of cereal weighs 750 g and costs £2.70.

Price per gram of cereal remains the same for each box size.

a) What would a 500 g box of the same cereal cost?

Show your method

£

b) How much would a 1.25 kg box of the same cereal cost?

Show your method

£

2 marks

Reasoning Test 3   Name _____

9 In a children's fantasy novel, a magic carpet shrinks every time it flies. The width reduces to a third of its previous size and the length reduces by half.

After **three flights**, the carpet's area is 4 m².

If the carpet's original width was 27 m what was its length?

Show your workings clearly.

Show your method

m

2 marks

10 In Jamaica, the total rainfall in June is 96 mm.

a) What would the average rainfall be for each day in June?

Show your method

ml

Reasoning Test 3

b) If the average daily rainfall in Jamaica is 2.5 ml, what is the total annual rainfall?

Give the answer in **litres**.

Assume this is NOT a leap year.

L

2 marks

11 The tallest building in the world is 830 m high and the highest mountain is 8,715 m high.

a) How many times higher is the tallest mountain than the tallest building?

Reasoning Test 3   Name _____

b) An aeroplane travels at a cruising altitude that is 15 times the height of the world's tallest building.

At what altitude does it fly? Give the answer in metres.

m

2 marks

12 A photobook company prints 50% of its books in December, 10% in each of the summer months of June, July and August and 2.5% in each of the remaining eight months.

a) If the company produces 24,635 books in March, how many books will it make in December?

b) If each book makes £3 profit, how much money does the company make in June?

3 marks

Total marks ..../18

Reasoning Test 4 Name _____

1 Choose the correct numbers to complete the calculations.

| 56 | | 70 | | 0.8 |

7 × 8 = ☐   7 × ☐ = 5.6   ☐ × 8 = 560

1 mark

2 This table shows the tide times at a local beach.

High/low tide	Time	Height
High tide	04:21	3.37 m
Low tide	10:32	0.87 m
High tide	16:44	3.38 m

How many minutes is it between high and low tide in the morning?

Show your method

☐ minutes

1 mark

3 Measure these two lengths.

What is the difference between the lengths? Give the answer in millimetres.

_____

_____

☐ mm

1 mark

Reasoning Test 4   Name _____

4 Tick the shapes that are nets of triangular prisms.

a)    b) 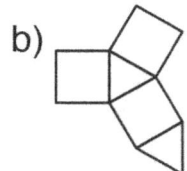   c) 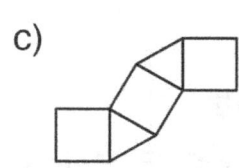   d)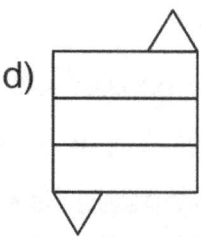

1 mark

5 Match the fractions that are equal.
One has been done for you.

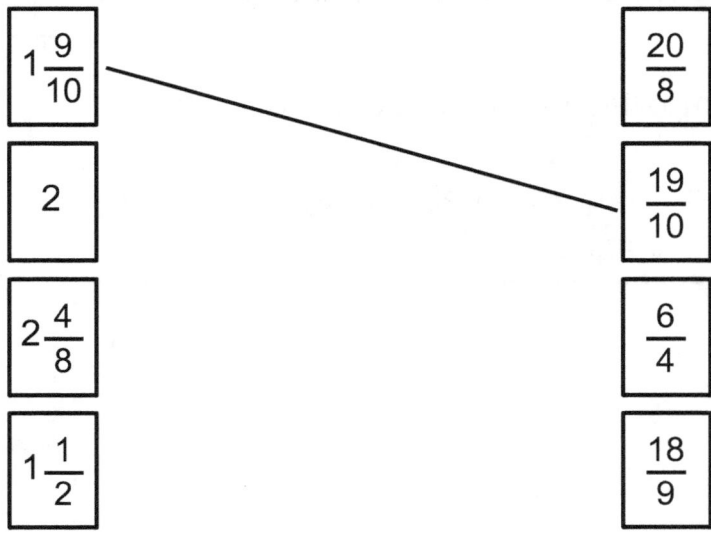

1 mark

6 Shade $\frac{6}{9}$ of each shape.

a) 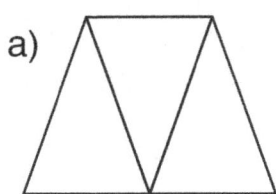   b) 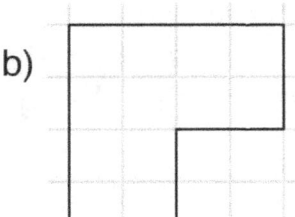   c)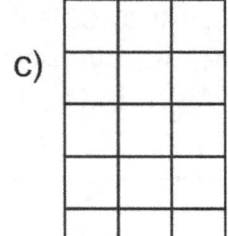

1 mark

Reasoning Test 4   Name _____

7 Jack and George both completed this calculation:

78 × 15

Jack's answer is **1,170**.

George's answer is **495**.

Whose answer is correct? Explain why.

[   ]

1 mark

8 Fifteen teams enter a cycling competition and there are 8 people in each team.

a) If each team member cycles 12 kilometres, what is the total distance covered by all the cyclists in the competition?

Show your method [   ] km

b) A third of the teams continue to the second part of the race but each team loses half its cyclists. The remaining cyclists go on to cycle a further 26 km each. What is the total distance that **this** group of cyclists covers in total?

Show your method [   ] km

2 marks

Reasoning Test 4   Name _____

9 a) Find the fourth term in the arithmetic sequence $5n - 3$.

Term: 1   2   3   4

Number: 2   7   12   ☐

b) What is the formula for this sequence?

Term: 1   2   3   4

Number: −2   4   10   16

Show your method

2 marks

10 Entry to a theme park in the winter months costs £7.80 per adult and £5.40 per child. In the summer months, these prices increase by 25%.

How much would it cost for a family of two adults and two children to enter the park during the summer?

Show your method

£

2 marks

Reasoning Test 4 Name _____

11 A box of chocolates has coffee cream and caramels in the ratio 3 : 5.

a) If there are 24 coffee creams in the box how many caramels will there be?

b) If the box costs £3.20, how much does each sweet cost?

p

2 marks

12 A radio DJ presents a breakfast show from 6.30 a.m. to 10 a.m. every weekday morning and a Sunday afternoon show between 1 p.m. and 4 p.m.

a) How many minutes per week is the DJ presenting shows on the radio?

minutes

Reasoning Test 4   Name _____

When off air, the DJ is expected to host three events a week in the local community. She gets paid £375 for each appearance.

b) How much would she earn if she worked for 28 weeks of the year hosting events?

3 marks

Total marks ..../18

Reasoning Test 5   Name _____

1 Circle three factors of 48.

    10   4   5   12   14   9   6

1 mark

2 This table shows the availability of a holiday cottage in July and August, 2019.

July, 2019 / August, 2019 calendars

Arrival date: 10/08/2019   Departure date: 17/08/2019

Book

☐ Free and possible arrival day
■ Free but day of arrival not allowed
☐ Full

What dates are available for arrival during the two-month period?

1 mark

3 Tick all the regular polygons.
Use a ruler to check lengths.

a) b) c) d) e) f)

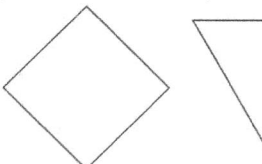

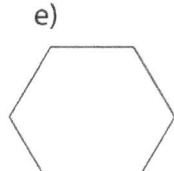

1 mark

Reasoning Test 5   Name _____

4 Enlarge this shape by a scale factor of 3. Write the coordinates of the vertices of the enlarged shape in the boxes below.

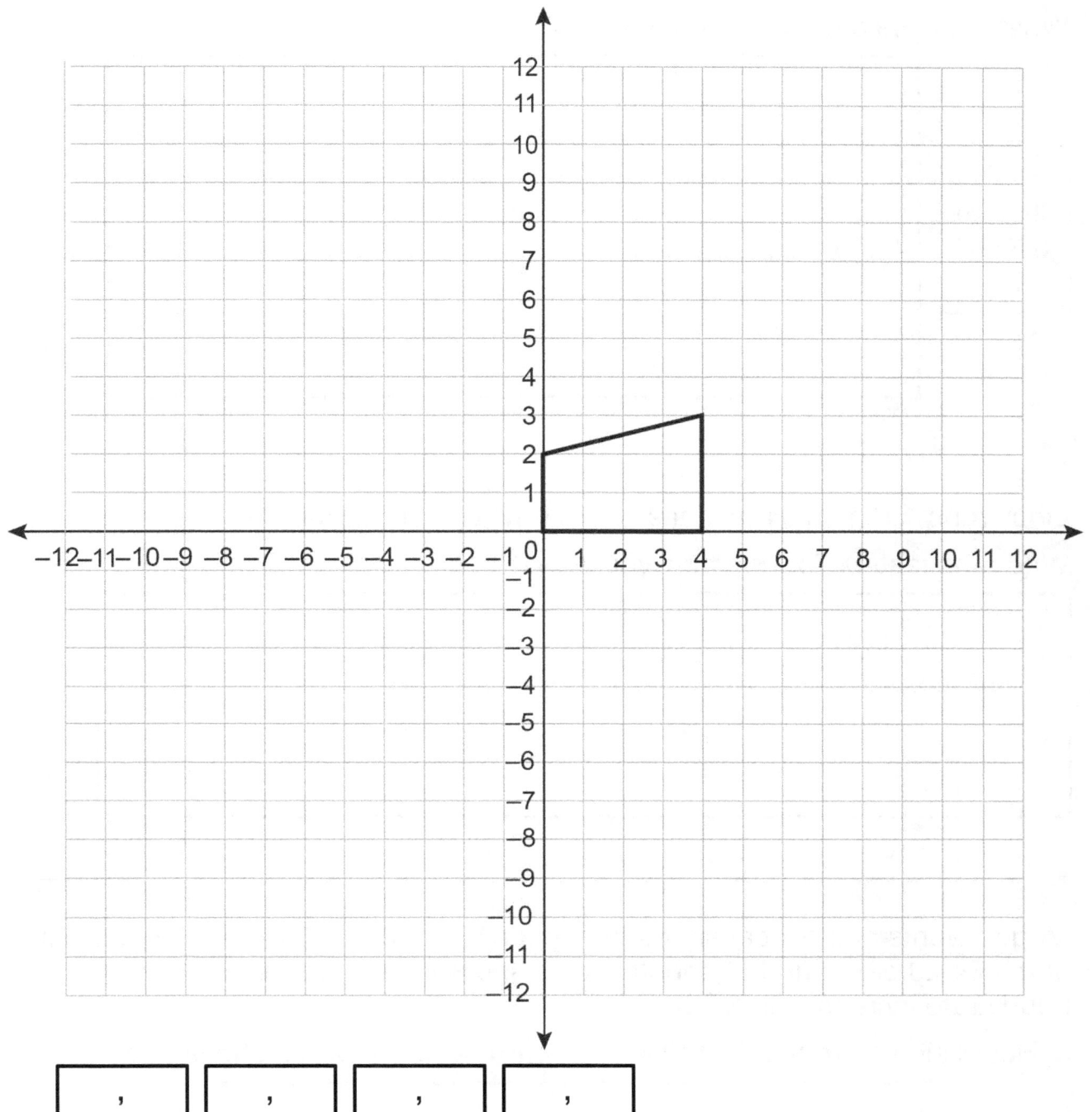

[,] [,] [,] [,]

1 mark

5 Add these two fractions.

$\frac{5}{20} + \frac{3}{5} =$ [   ]

1 mark

Reasoning Test 5

Name _____

6 Last week, a television cost £450. This week, it is in the sale with a 30% reduction on the price.

What is the sale price of the television?

Show your method

£ _____

1 mark

7 Libby runs 10 km in 54 minutes. Arjun runs 8 km in 40 minutes.

Who runs faster? Explain how you know.

1 mark

8 Magda works at the local sports centre for 16 hours each week. She currently earns £4.20 per hour. In 2 months' time she will be 18 years old. Then her hourly rate increases to £5.90.

a) How much more money will Magda earn per week when she is 18?

Show your method

£ _____

b) If Laura was making seven chocolate pots, how much milk would she need?

Show your method

ml

2 marks

12 On a long car journey, Anja and Finlay count different-coloured cars. In 150 seconds, they see 5 red cars.

a) How many red cars will they see, on average, in 30 minutes?

90 seconds = 1.5 minutes

Show your method

Reasoning Test 5   Name _____

They also count 7 white cars and 2 black cars every 90 seconds.

b) How many white cars and black cars will they see in 30 minutes?

Show your method

3 marks

Total marks ..../18

Reasoning Test 6

Name _____

1. Choose two numbers from the number cards below to complete the sequence.

    1, 3, 6, 10, 15, ___, ___

    | 20 | 25 | 18 | 21 | 30 | 28 | 24 |

    1 mark

2. This pie chart shows 200 people's favourite types of film.

    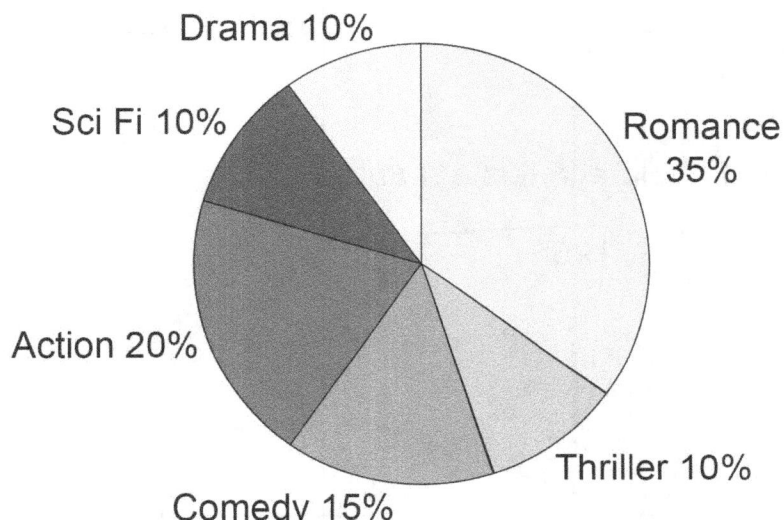

    Favourite film

    a) What percentage of the people surveyed liked Romance?

    _____ %

    b) How many people chose Comedy as their favourite type of film?

    _____ people

    1 mark

Reasoning Test 6   Name _____

3 Look at the shapes and complete the table.
One has been done for you.

a) 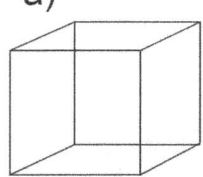   b) 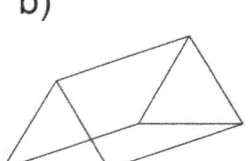   c) 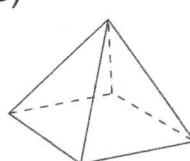   d)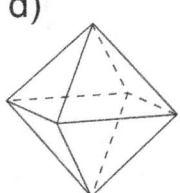

Shape	Faces	Edges	Vertices
Cube	6	12	8
Triangular prism			
Square-based pyramid			
Octahedron			

1 mark

4 What is the size of angle *a* in this quadrilateral?

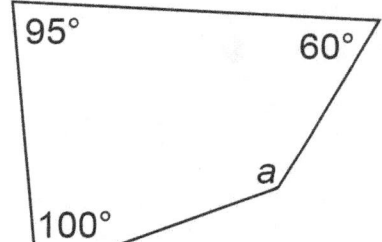

Show your method

_____°

1 mark

Reasoning Test 6   Name _____

5 Which is greater, $\frac{4}{5}$ of 120 or $\frac{3}{5}$ of 150?

Show your method

1 mark

6 Order these fractions on the number line below.

$$\frac{2}{8}, \frac{3}{4}, \frac{9}{18}, \frac{4}{2}, \frac{8}{8}$$

Convert the fractions before ordering them and use a ruler to ensure your answer is accurate.

0   1   2

Show your method

1 mark

Reasoning Test 6

Name _____

7 Rainfall is measured in two towns over a 24-hour period.

Location (town)	Rainfall per hour (mm)	Hours of rain (in 24 hours)
Larwick	5.8	4
Mount-Harford	2.5	12

In which town did more rain fall?

Explain how you know.

1 mark

8 Larina goes to watch a film at the cinema. The film lasts for 2 hours 45 minutes.

a) If the film finishes at 7.25 p.m., what time did it start?

Show your method

_____ p.m.

It takes Larina 45 minutes to walk home.

b) If she leaves at the end of the film, what time will she arrive home?

Show your method

_____ p.m.

2 marks

Reasoning Test 6

Name _____

9 Think of a number.
- Multiply it by two.
- Add ten.
- Halve the answer.
- Subtract your original number.

Why will you always be left with 5?

2 marks

10 A toy shop orders 40 boxes of bears. In each box there are 4 bags, each containing 8 bears.

a) How many bears are there in the 40 boxes?

bears

Reasoning Test 6   Name _____

Each box of bears costs £18.

b) How much does the whole order cost the toy shop?

£ _____

2 marks

11 On a farm, there are 135 cows, three times as many sheep and one-fifth as many ducks as there are cows.

a) How many ducks and sheep are there on the farm?

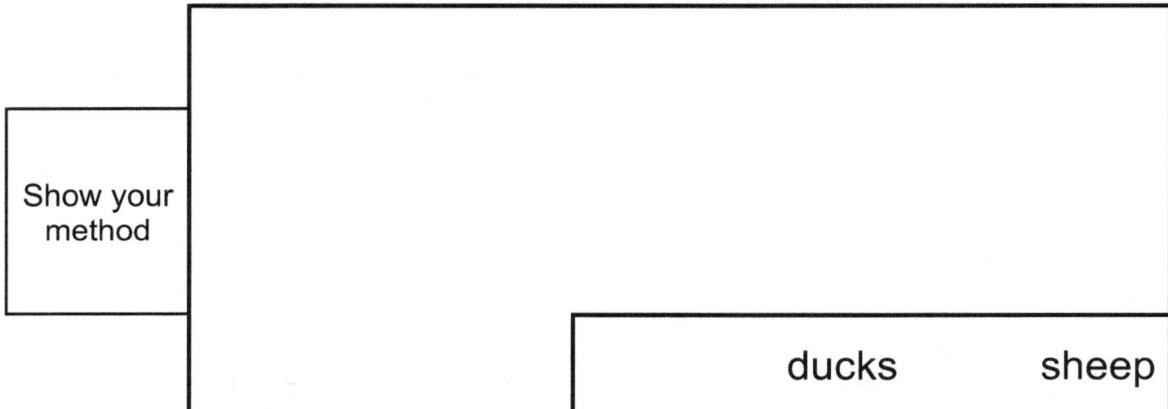

_____ ducks   _____ sheep

b) The farm measures 20 km².

What are two possible lengths and widths of the farm boundary?

_____ km × _____ km

2 marks

Reasoning Test 6   Name _____

12 At the weekend, Joshua's grandma gave him some sweets.

On Sunday he ate half of them plus one more.

On Monday he ate half of the rest plus one more.

On Tuesday he ate half of the rest plus one more again.

On Wednesday he found there was only one left.

How many sweets did Joshua's grandma give him at the weekend?

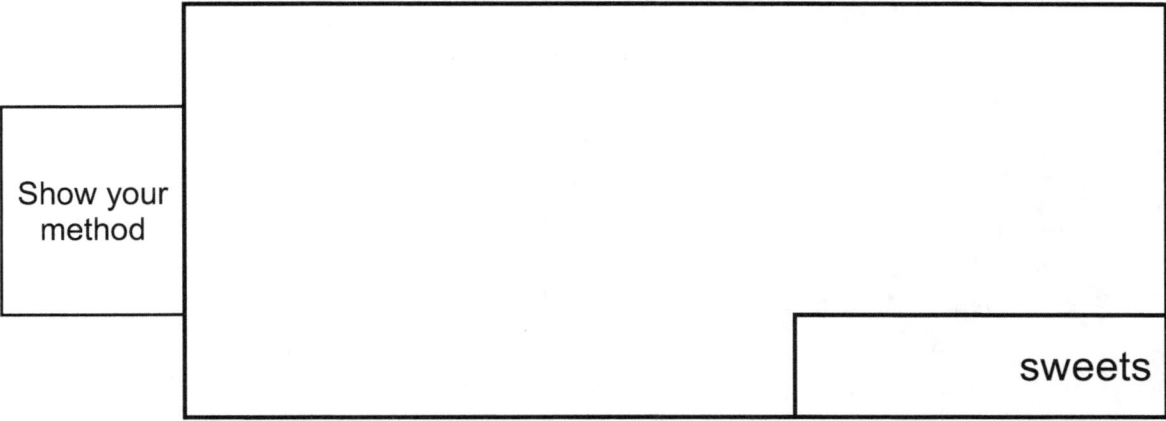

_____ sweets

3 marks

Total marks ………/18

Reasoning Test 7

1. Write down two numbers from the sequence that add to make a total of 72.

   | 8 | 16 | 24 | 32 | 40 | 48 | 56 |

   1 mark

2. This chart shows the favourite snacks of children in Year 6.

   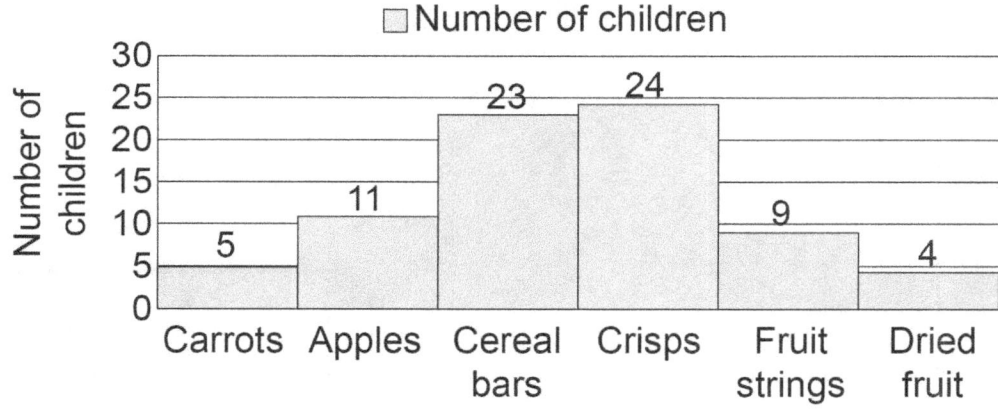

   a) How many more children liked crisps than liked apples?

   b) How many children were included in the data collection?

   1 mark

3. In the space below, draw a horizontal line 58 mm long.

   1 mark

4 This shape is made up of four cubes stacked together.

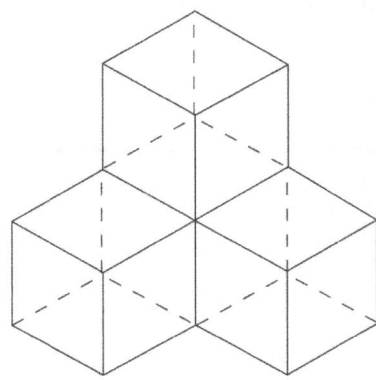

If you could pick this shape up and turn it around so you could count all the faces, how many faces would you be able to see in total?

1 mark

5 Multiply these two fractions together.

$$\frac{5}{8} \times \frac{1}{5} =$$

1 mark

Reasoning Test 7

6 Select two fractions that are equal to $\frac{2}{5}$.

$\frac{40}{100}$ $\frac{1}{5}$ $\frac{4}{10}$ 0.4 $\frac{8}{10}$ $\frac{16}{40}$

Show your method

1 mark

7 Billy and Declan are discussing prime numbers. Billy says that 17 is not a prime number. Declan says he is wrong.

Who is correct? Explain why.

1 mark

8 Golf balls come in packs of six. Each pack costs £12.00.

Rob needs 80 golf balls for a competition he is organising. How much will it cost him?

Show your method

£

2 marks

Reasoning Test 7   Name _____

9   Grace thinks of a whole number.

She divides it by four to make another whole number.

She rounds her answer to the nearest 10.

Her answer is 20.

Give two possible numbers that Grace could have started with.

Show your method

2 marks

10 Duvets for beds come in single, double and king size.

A single duvet is 180 cm long × 135 cm wide.

a) If the width of a double duvet is $\frac{1}{3}$ more than the width of a single duvet, what will its measurements be?

Show your method

cm ×   cm

© HarperCollins*Publishers* Ltd 2019

Reasoning Test 7

b) The width of a king size duvet is $\frac{1}{3}$ more again than the width of a double duvet. What will its measurements be?

cm × cm

2 marks

11 In a school, the proportion of boys to girls is 3 in 5.

a) If there are 36 boys, how many girls are there in the school?

girls

b) If the number of girls was 100, how many boys would there be?

boys

2 marks

Reasoning Test 7   Name _____

12 A delivery lorry leaves the depot at 7.30 a.m. It makes 17 stops. On average, it makes one stop every 10 minutes.

a) What time will the driver complete the deliveries?

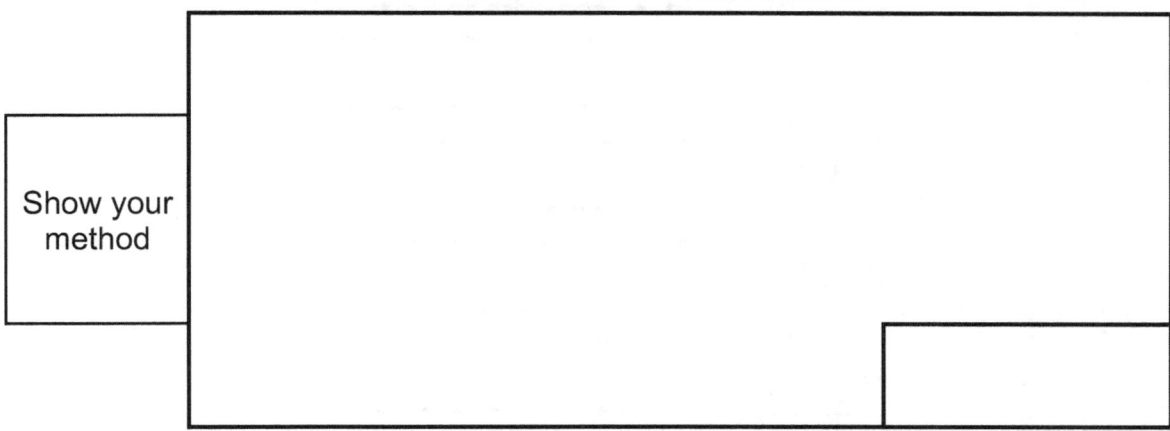

Show your method

The delivery driver gets a £2.60 bonus for each delivery made on time.

b) If all 17 deliveries are made on time, how much could be earned in bonuses on this day?

Show your method

£

3 marks

Total marks ………/18

Reasoning Test 8   Name _____

1 Round each of these decimals to the nearest whole number.

11.8	
22.1	
0.85	
125.2	
66.7	

1 mark

2 This is the nutritional content of almond milk.

a) How much fibre would there be in 200 ml?

b) How much milk would you have to drink to consume 65 calories?

Nutrition Typical values	per 100 ml
Energy	53 kJ/13 calories
Fat	1.1 g
of which	
• saturates	0.1 g
Carbohydrates	nil
of which	
• sugars	nil
Fibre	0.4 g
Protein	0.4 g
Salt	0.13 g
Vitamins:	
• D	0.75 µg*
• E	1.80 mg*
• Riboflavin (B2)	0.21 mg
• B12	0.38 µg*
Minerals:	
• Calcium	120 mg

* = 15% of the nutrient values

These values are approximate due to the variations that occur in natural ingredients.

1 mark

Reasoning Test 8   Name _____

3 Circle all the obtuse angles on the shapes below.

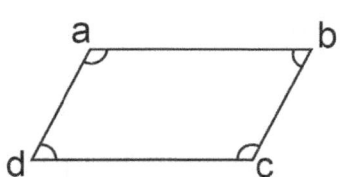 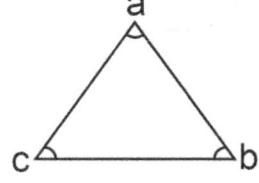 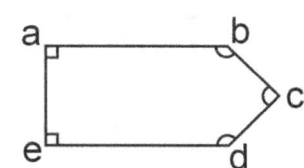

1 mark

4 Rotate this shape 180° around the point (0, 0) and draw its new position on the grid.

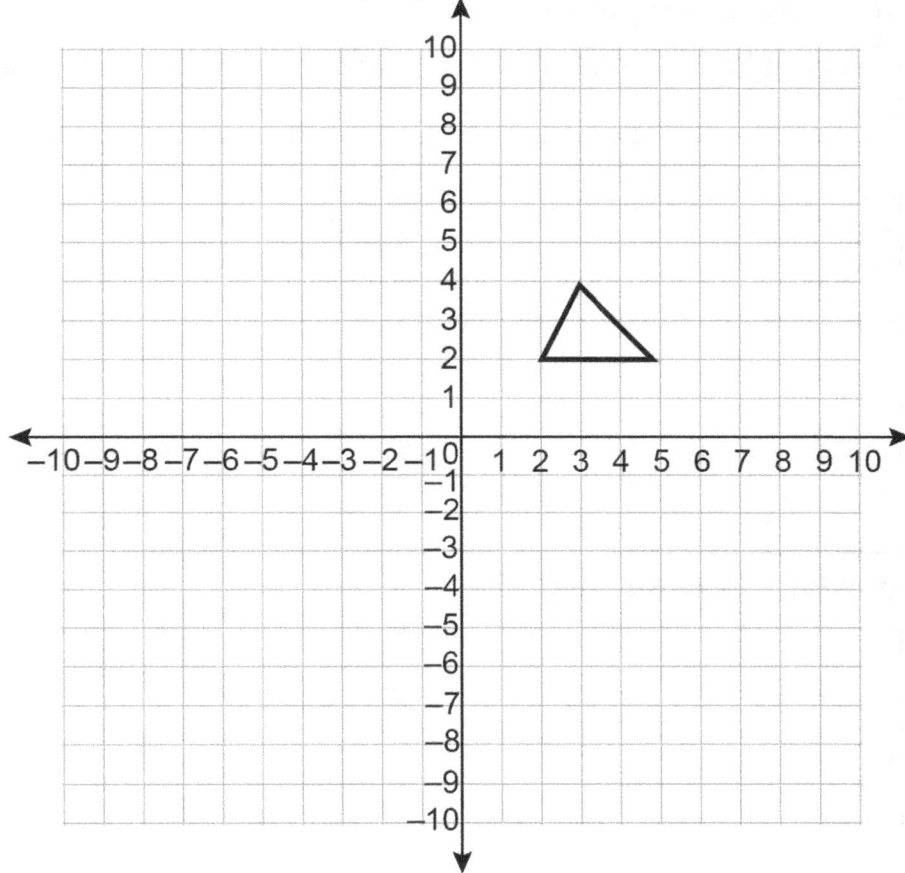

1 mark

5 Add these fractions and write your answer as a mixed number.

$\frac{1}{2} + \frac{3}{6} + \frac{8}{12} =$ ☐

1 mark

6 Add these decimal fractions and write your answer as a mixed number.

0.1 + 0.8 + 1.3 = ☐

1 mark

Reasoning Test 8

Name _____

7 If 16 is a square number, why is 32 not?

Explain your answer in the space below.

1 mark

8 Madeline's mum pays £86 per term for her flute lessons.

a) If the lessons cost £10.75 each, how many lessons does she have each term?

Show your method

lessons

b) How much do 15 lessons cost?

Show your method

£

2 marks

Reasoning Test 8

9 This year, Harry's dad is 8 times as old as Harry.

a) If he is 40 years old, how old is Harry?

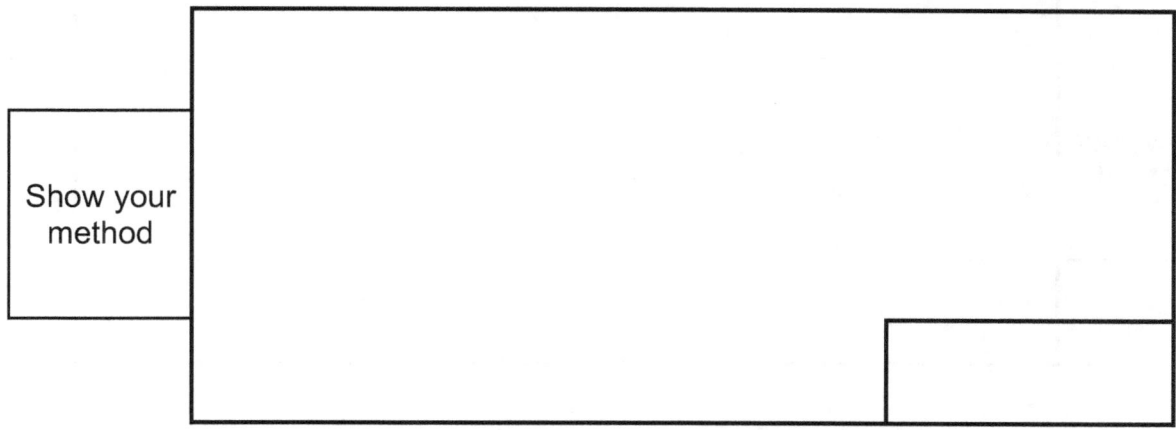

b) When Harry is 25 years old, what age will his dad be?

2 marks

10 One of the longest snakes in the world is 7.8 m long. The shortest snake is 10 cm long.

a) How many times as long as the shortest snake is the longest?

Reasoning Test 8   Name _____

b) What would the total length of five of the longest snakes be?

m

2 marks

11 Sophie's dad says that for every £2 she saves, he will give her another 40%.

a) If Sophie saves £15, how much money will her dad give her?

£

b) If Sophie's dad gives her £10, how much has she saved?

£

2 marks

Reasoning Test 8

Name _____

12 A family is buying a new house.

They want to build a fence all the way round the property.

This is an outline of the property.

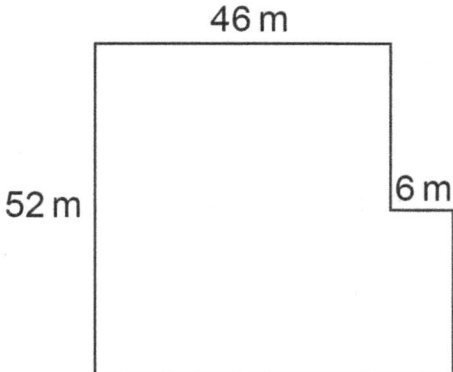

a) How many metres of fence do they need to buy?

The fence panels come in packs of 3 m and cost £17.50 per pack.

b) How much will it cost to fence the boundary of the property?

3 marks

Total marks ………/18

Reasoning Test 9   Name _____

1. Choose the correct number from those below and write it in the middle box to complete this calculation.

   | 55 | × | | = | 27.5 |

   25%   0.05   27.5   0.5 $\frac{2}{5}$

   1 mark

2. The first clock shows the time that Veejay goes to work. After 3.5 hours, he has a break.

    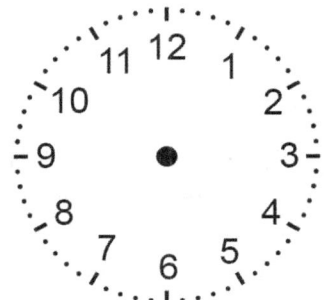

   a) Draw the time on the second clock to show when Veejay has his break.
   b) Write the time that Veejay has his break.

   | : |

   1 mark

3. Draw and label a square with an area of 36 cm².
   Use a ruler to answer this question.

   1 mark

Reasoning Test 9   Name _____

4 Label the shapes that you can see in the tangram.

   One has been done for you.

   You can use more than one shape to make some of the different possible shapes.

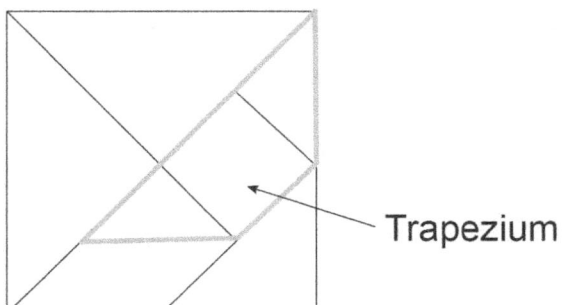

Trapezium	✓
Triangle	
Parallelogram	
Square	
Pentagon	

1 mark

5 Put each of these calculations in the correct box.

   One has been done for you.

   $\frac{1}{4} \times 5$ $\frac{1}{4} \times 6$ $\frac{1}{3} \times 2$

   $\frac{1}{8} \times 8$ $\frac{1}{6} \times 3$ $\frac{1}{9} \times 10$

Less than 1	Equal to 1	Greater than 1
$\frac{1}{6} \times 3$		

1 mark

6 Which calculation gives the greatest answer?

   Circle the correct box.

   0.4×5 $\frac{3}{10}$ of 20   25% of 8

1 mark

Reasoning Test 9

Name _____

7 Sam and Sid have arranged to meet at Budapest Airport.

Sam takes off from London for Budapest at 17:20. The flight takes 2 hours 25 minutes.

Sid takes off from Edinburgh for Budapest at 16.45. The flight takes 3 hours 5 minutes.

Who will arrive at Budapest airport first?

Show your method

_____ arrives first

1 mark

8 An ice-cream factory produces 125 litres of ice cream every minute.

a) How much ice cream is produced in 10 minutes?

Show your method

_____ litres

b) How long will it take the factory to make 2,500 litres of ice cream?

Show your method

minutes

2 marks

9 a) What is the value of x?

Show your method

x =

b) What is the value of x in this equation?

$$\frac{4x + 5}{9} = 5$$

x =

2 marks

10 A jeweller ordered 12 boxes of plastic beads. There were 655 beads in each box. How many beads were there in the order?

Show your method

beads

2 marks

11 Tristan works at a hairdressers. Every three months, he gets a 20% bonus on his income. At the end of March, his bonus was £600.

How much money did he earn from January to March, before his bonus?

Show your method

£

2 marks

12 On Earth, there are 365 days each year and 366 days in a leap year (every fourth year).

a) How many days are there in 4 consecutive years on Earth?

days

On Mars, there are 687 days in a year.

b) How many weeks are there in one year on Mars?

Round your answer to the nearest whole number when calculating weeks on Mars.

weeks

3 marks

Total marks ………/18

Reasoning Test 10   Name _____

1 Choose the correct number from those below and write it in the middle box to complete each calculation.

| 67 | × | | = | 670 |

| 67 | × | | = | 6,700 |

| 67 | × | | = | 6.7 |

| 10 | 100 | 1,000 | 0.01 | 1.1 | 0.1 |

1 mark

2 This pictogram shows the lifespans of some animals.

Rabbit	● ● ● ● ●
Cat	● ● ● ● ● ● ● ●
Mouse	●
Chicken	● ● ● ●
Horse	● ● ● ● ● ● ● ● ● ● ● ● ● ●

2 years = ●   1 year = ●

How many more years does a horse live than a rabbit?

| years |

1 mark

3 Calculate the volume of this cuboid.

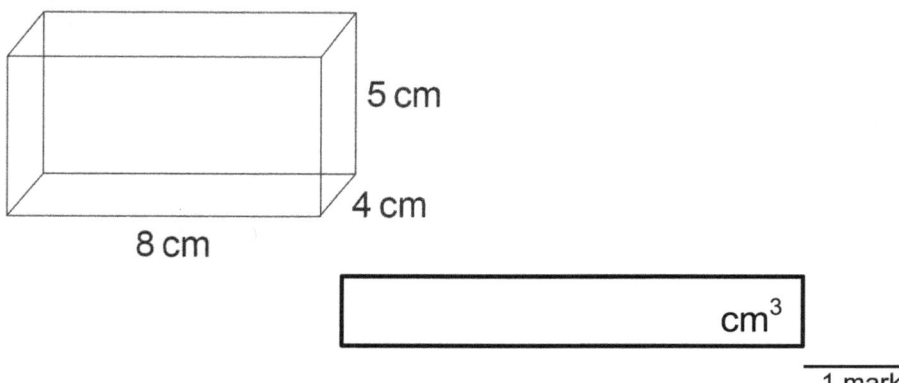

5 cm
4 cm
8 cm

| cm³ |

1 mark

Reasoning Test 10   Name _____

4 Calculate the unknown angle in this shape.

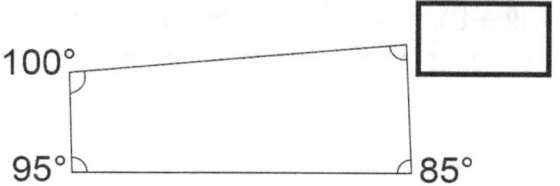

1 mark

5 Choose the correct sign to make this equation correct.

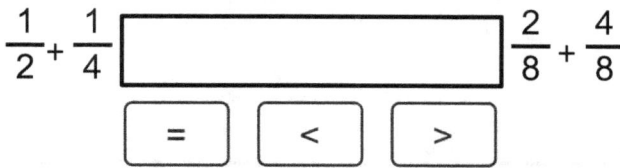

1 mark

6 Divide each of these shapes and then shade the correct fraction.

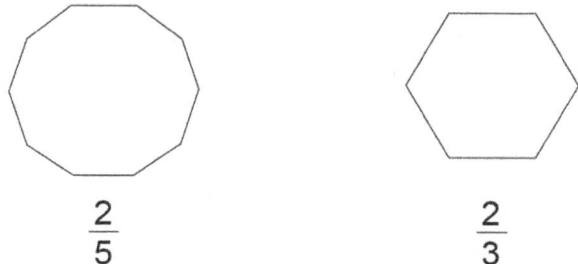

1 mark

7 Circle the amount that is worth more.

Explain your answer.

$\frac{5}{8}$ of £136   or $\frac{7}{9}$ of £171

1 mark

Reasoning Test 10

Name _____

8 Jasmine runs a bath. It takes 10 minutes to run 60 litres of water into the bath.

a) How long does it take to run 12 litres into the same bath?

Show your method

_____ minutes

b) If the water ran at twice the speed, how long would it take to run 132 litres?

Show your method

_____ minutes

2 marks

9 Jamille thinks of a number.

He says his number is square, it is odd and is between 50 and 100.

What is his number? Explain how you know.

Show your method

2 marks

Reasoning Test 10   Name _____

10 A train departs from the station at 8:44 a.m.. It arrives at its destination at 11:04 a.m..

a) How long does the journey take?

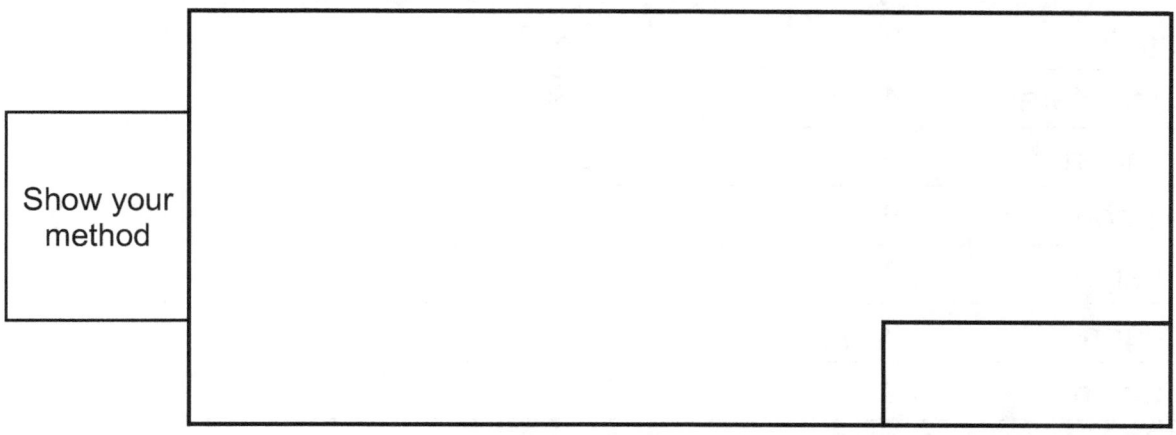

b) If the train is travelling at an average speed of 90 mph, what is the total distance for the journey?

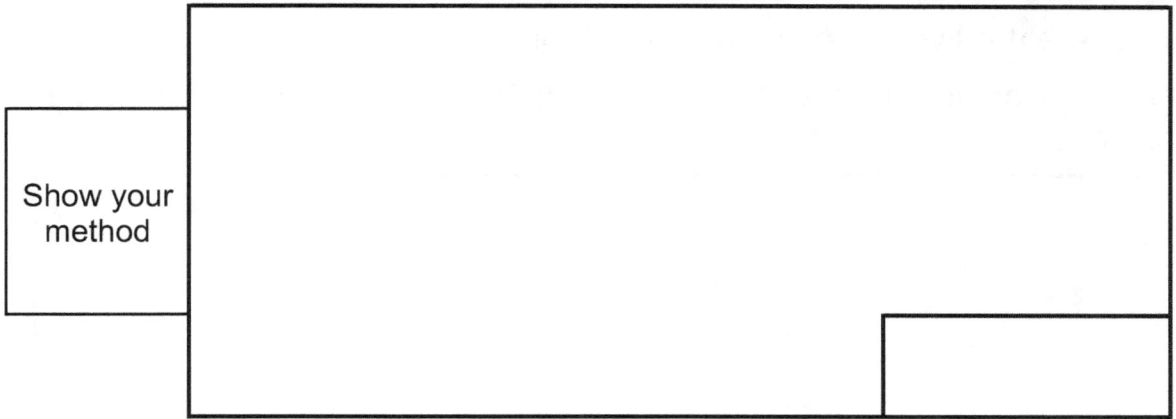

2 marks

11 A theatre production company has 650 tickets to sell for its latest play. The local theatre that hosts the play receives 30% of all tickets sales.

If the tickets cost £9 each and all tickets are sold, what is the maximum amount of money the local theatre makes from ticket sales?

£

2 marks

Reasoning Test 10

Name _____

12 The sum of the internal angles of any polygon increases by 180° every time another side is added.

Shape	No. of sides (n)	Sum of internal angles
Triangle	3	180°
Quadrilateral	4	360°
Pentagon	5	540°
Hexagon	6	
Heptagon	7	
Octagon	8	
Nonagon	9	
Decagon	10	
Any shape	$n - 2$	

a) Complete the five empty rows of the table.

b) Now write down a rule for finding the sum of the internal angles of any polygon.

3 marks

Total marks ..../18

Reasoning Test 11   Name _____

1 Circle all the multiples of 3.

   100   23   72   3   600   18

   _____
   1 mark

2 Children in Class 2 measured their heart rate immediately after two minutes of exercise.

Group One	Heart rate (beats per 30 seconds)
Ben	80
Jack	72
Millie	68
Amy	84
George	77

a) What is Millie's heart rate per minute?

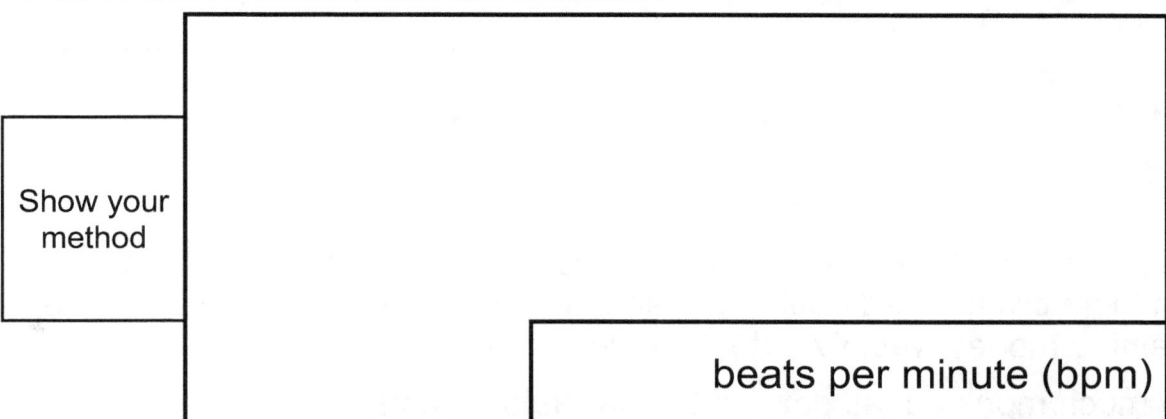

beats per minute (bpm)

b) What is the difference between the fastest and slowest heart rates, in bpm?

bpm

_____
1 mark

Reasoning Test 11   Name _____

3   How many single cubes make up this shape?

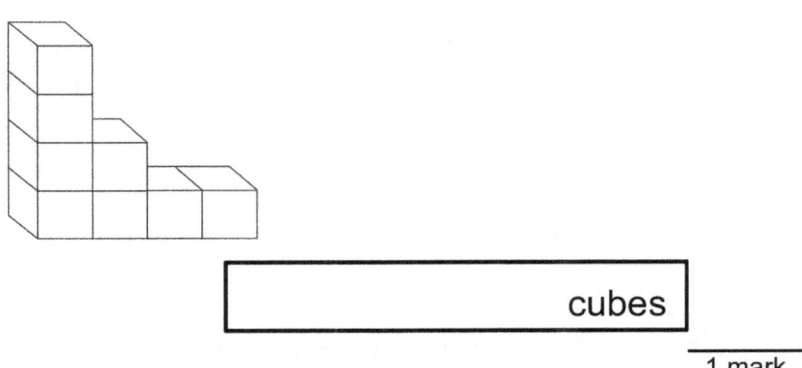

                                         [   ] cubes
                                              1 mark

4   This is a cuboid.
    Calculate the total length of all its edges.

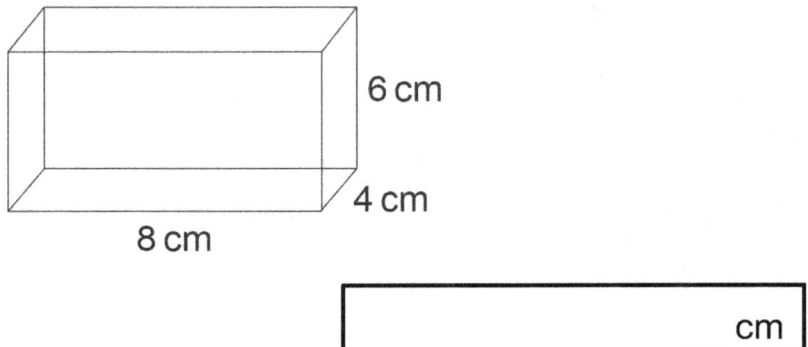

                                         [   ] cm
                                              1 mark

5   Three brothers shared £40. The youngest brother received $\frac{1}{5}$ of the money, the middle brother had double the amount given to the youngest and the remaining money was given to the eldest brother.

How much money does each of the brothers receive?

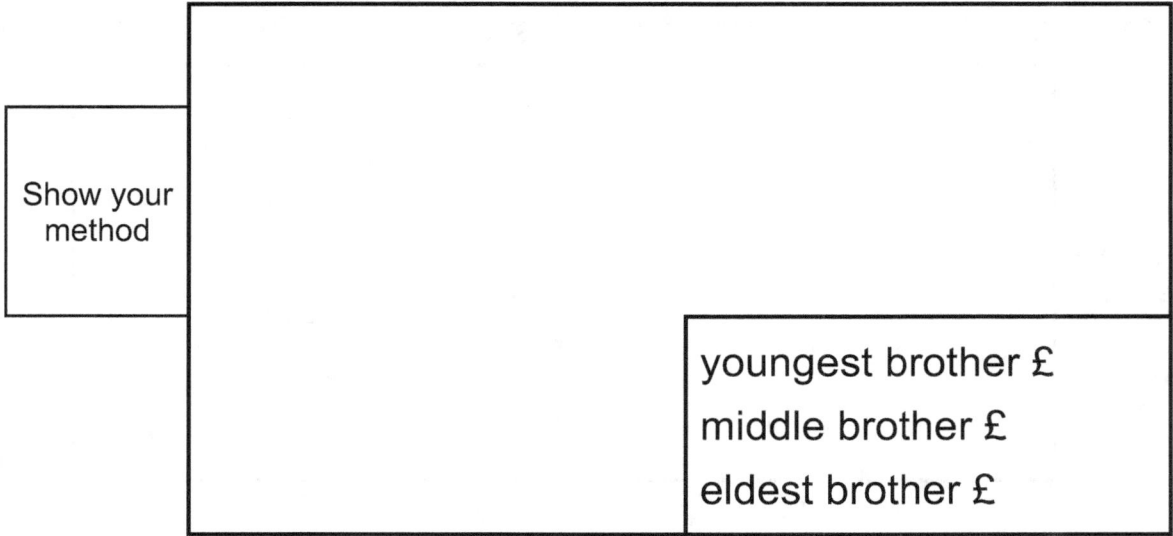

1 mark

6 In each number card, circle the smaller amount.

| 9.4   9 5/100 | | 0.45   42/100 | | 2.5   2 5/8 |

1 mark

7 Katie has three silver coins and one copper coin in her pocket.
Can Katie have more than £2 in her pocket? Explain your answer.

Yes / No

1 mark

8 A school receives a delivery of 86 boxes of text books. There are 24 books in each box.

a) How many books are there altogether in the delivery?

Show your method

books

Reasoning Test 11

Name _____

b) The books are divided equally among 8 classes. How many books does each class receive?

Show your method

_____ books

2 marks

9 Children in Year 6 are measuring their heights.

| Seb | Jules | Cassius | Isla | Ava |
| 150 cm | 145 cm | 148 cm | 140 cm | 138 cm |

Amaya works out their average height to be 142 cm. Is she correct?

Show your method

2 marks

Reasoning Test 11

Name _____

10 A Formula One racing driver completes a 72-lap race.

On average, each lap takes 1 minute 15 seconds.

How long does it take for the driver to complete the race?

minutes

2 marks

11 In a sale, the price of a games console has been reduced by 20%. It now costs £98.

What was its original price?

£

2 marks

Reasoning Test 11

Name _____

12 At the summer fayre, Mrs White runs a tombola stall. Each ticket costs 75p. The prizes for the tombola cost £18 altogether.

a) How many tickets does Mrs White need to sell before she starts making a profit?

When Mrs White counts her money at the end of the fayre, she has £55.50 in her pot.

b) How much profit has she made?

3 marks

Total marks ………/18

Reasoning Test 12   Name _____

1 Circle the value that this number written in Roman numerals represents.

CCL

| 25 | 220 | 202 | 250 | 110 |

1 mark

2 Year 6 collected data from 340 children across the school about their family pets.

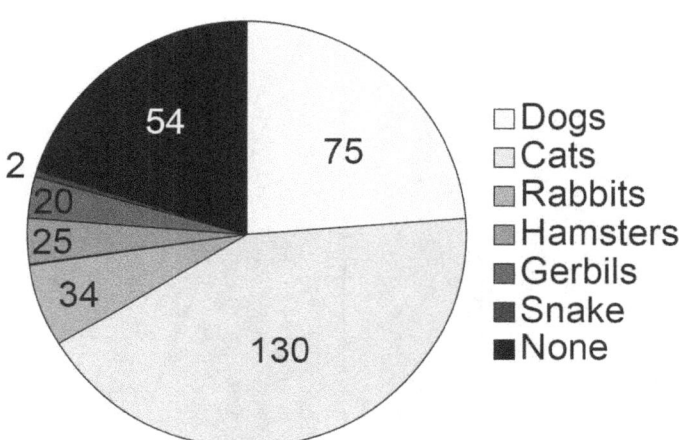

Family Pets in our School
- Dogs
- Cats
- Rabbits
- Hamsters
- Gerbils
- Snake
- None

a) What percentage of children had a rabbit?

[ ]%

b) How many more children had cats than had no pets at all?

[ ]

1 mark

Reasoning Test 12   Name _____

3 Calculate the diameter of this circle.

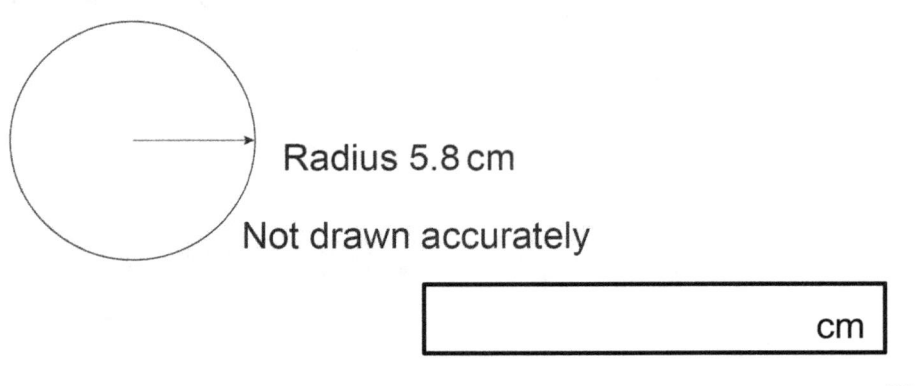

Radius 5.8 cm

Not drawn accurately

[_____ cm]

1 mark

4 Measure the dimensions of this triangle and calculate its area.

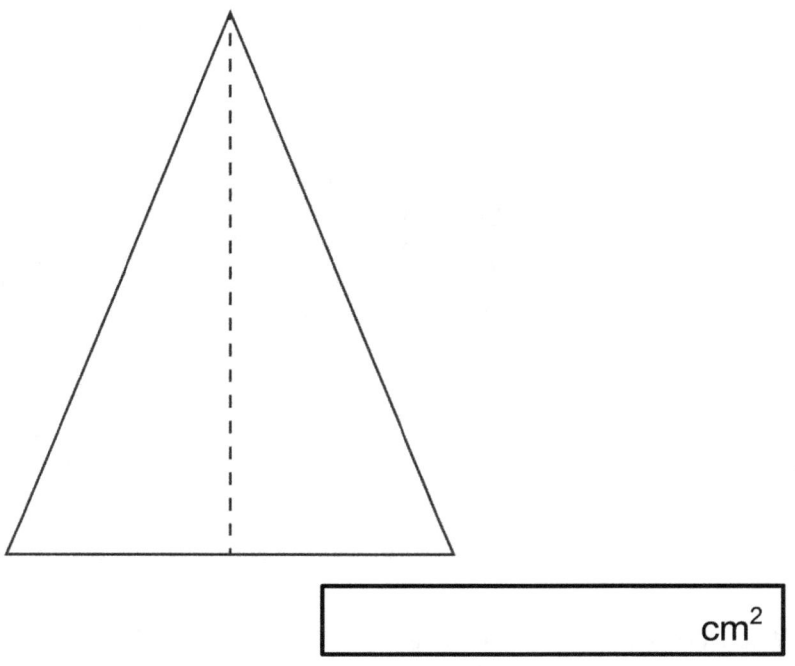

[_____ cm²]

1 mark

5 Calculate $\frac{2}{9}$ of 630.

Show your method

1 mark

Reasoning Test 12   Name _____

6 Complete this calculation. Write your answer as a mixed number.

$\dfrac{5}{6} + 2\dfrac{1}{3} + 1\dfrac{1}{2} + \dfrac{7}{6} =$

1 mark

7 'The sum of two negative numbers is always negative.'

Is this statement true, or false?

Explain your answer.

Show your method	
	True / False

1 mark

8 a) It takes 385 water drops to fill a 200 ml beaker. How many drops will it take for the water to reach the 40 ml mark on the beaker?

Show your method	
	drops

Reasoning Test 12

Name _____

b) If it takes an hour to fill the 200 ml beaker, how long would it take to fill a 750 ml beaker?

Give the answer in hours and minutes.

Show your method

2 marks

9 Thirty friends attend a party. They all shake each other's hands on arrival.

How many handshakes are made in total at the party?

Show your method

_____ handshakes

2 marks

10 The area of this garden is 144 m².

84 m² is covered in grass.

What fraction of the garden is not covered in grass?

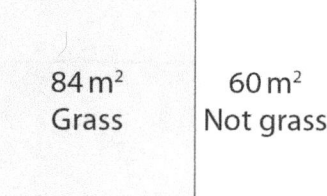

Area of garden = 144 m²
(Not drawn accurately)

| 84 m² Grass | 60 m² Not grass |

2 marks

Reasoning Test 12

Name _____

11 If 2.5 m of material costs £3.50, how much would 3.5 metres cost?

2 marks

12 The temperature at the base of Mount Snowdon is 23.1 °C and the temperature at the summit is 7.7 °C.

a) What is the difference in temperature between the base and the summit?

b) How many times as warm is it at the base than the summit?

3 marks

Total marks ………/18

Reasoning Test 13

Name _____

1 Tick the numbers that round to 170 when rounded to the nearest ten.

| 175 | 165 | 168 | 178 | 162 | 145 |

1 mark

2 A bus arrives at Stop three 56 minutes after leaving the bus station.
What time does it arrive at Stop three?
Complete the table.

Departs bus station	11:47
Stop one	12:03
Stop two	12:17
Stop three	
Final destination	1:02

1 mark

3 Draw an obtuse angle.

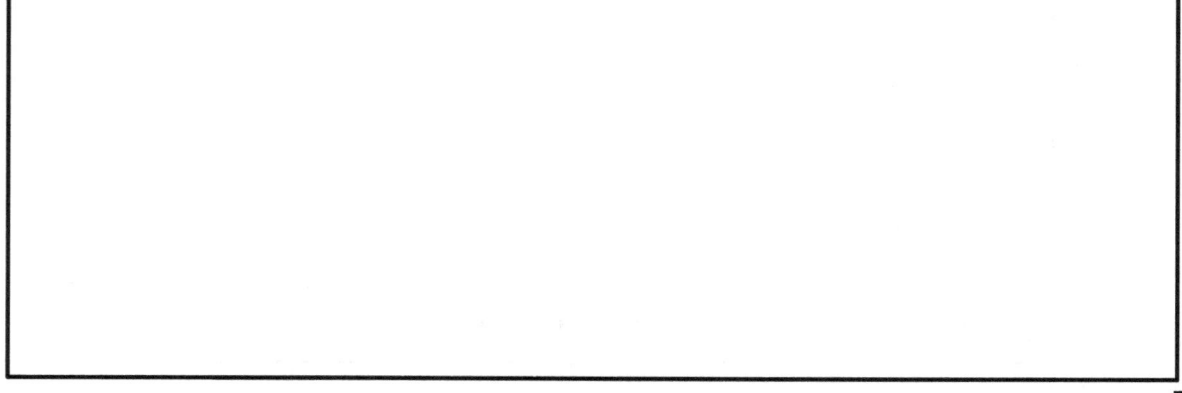

1 mark

4 Reflect this shape in the mirror line.

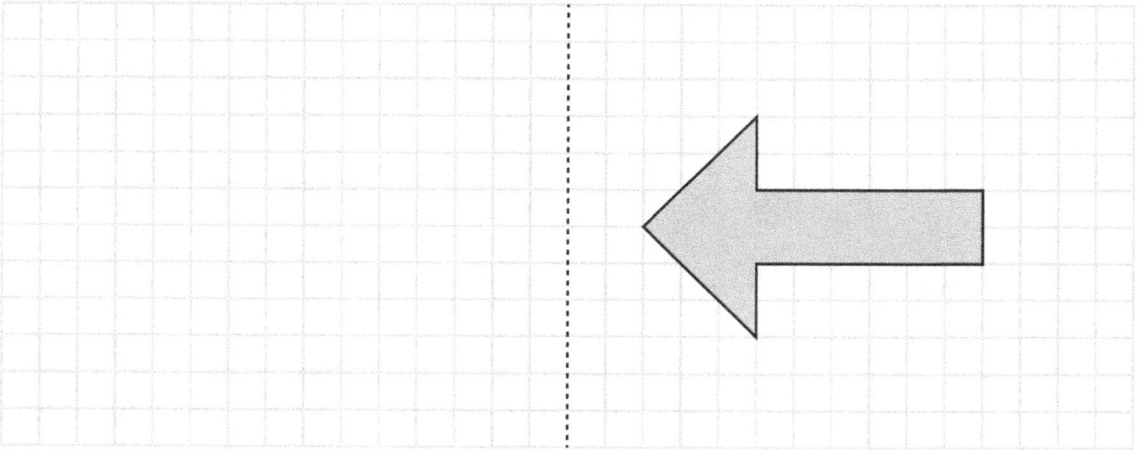

1 mark

Reasoning Test 13   Name _____

5 Add these fractions.

Give your answer as a mixed number.

$1\frac{1}{2} + 2\frac{3}{8} + \frac{5}{4} =$

Show your method

1 mark

6 Complete this calculation. Write your answer as a decimal number.

0.2 + 4.53 + 1.2 =

1 mark

7 Which is greater, 4 pints or 2.2 litres?
Explain your answer.

There are 568 ml in 1 pint.

Show your method

1 mark

Reasoning Test 13

Name _____

8 A train is travelling at an average speed of 70 mph.

If the journey takes 4.5 hours, how many miles will the train travel altogether?

Show your method

_____ miles

2 marks

9 a) Write the missing numbers in the table.

b) What is the rule for this sequence? Write this in the n column.

Term	1	2	3	4	5	6	8	10	n
No. of dots	5	7	9	11	13	15			

2 marks

Reasoning Test 13   Name _____

10 A window fitter is buying glass for windows in a new house. He needs to order glass for 8 windows, each with these dimensions.

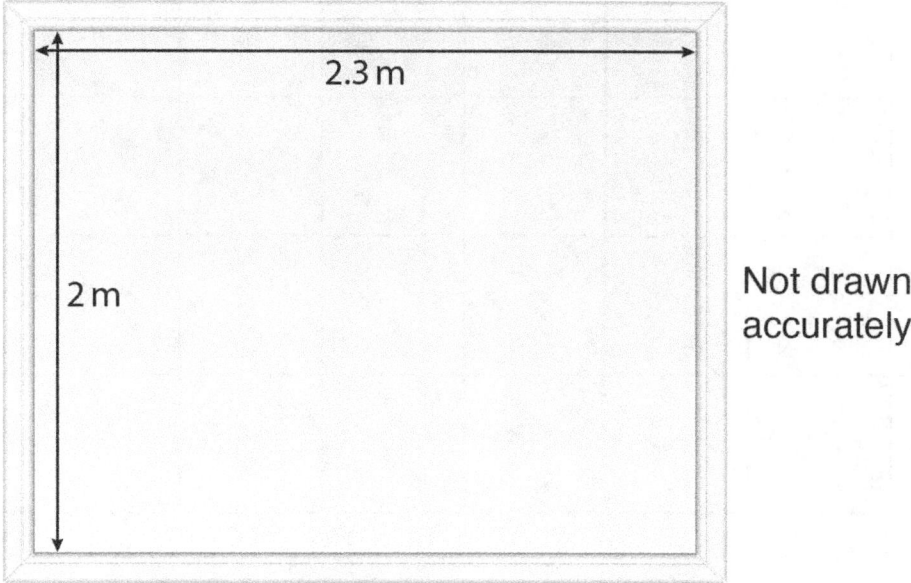

Not drawn accurately

What is the total area of glass that the window fitter needs to order?

Show your method

m²

2 marks

11 a) What is the ratio of grey slabs to white slabs in this patio?

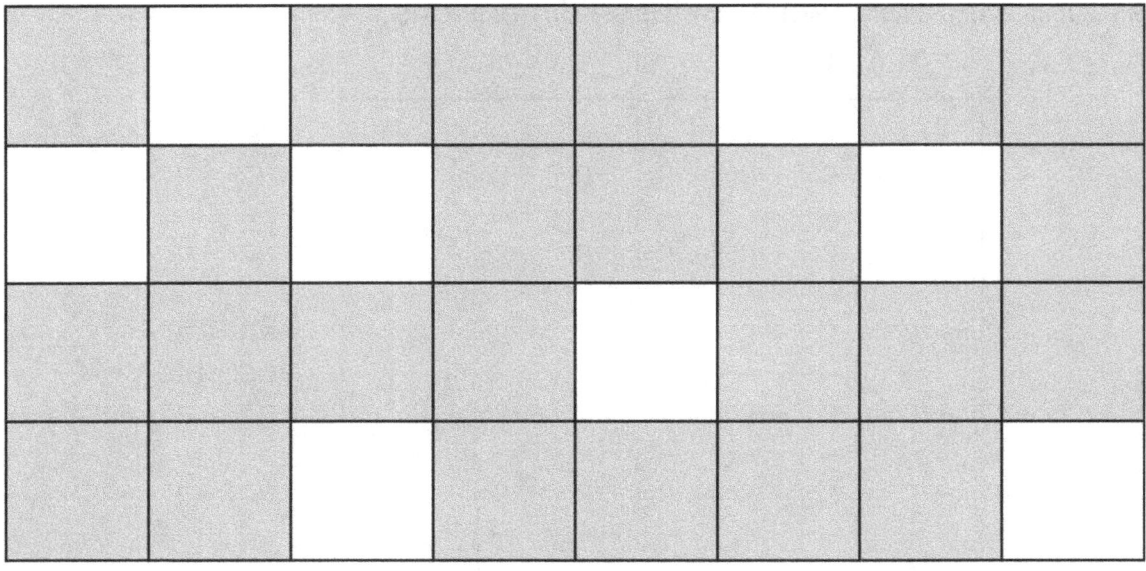

Not drawn accurately

Show your method

b) The patio is extended, keeping the colours in the same ratio. If 48 white slabs were used, how many grey slabs would be needed?

Show your method

slabs

2 marks

Reasoning Test 13   Name _____

12 A car is in a sale for £4,500 after its price was reduced by 40%.

a) What was the cost of the car before its price was reduced?

A monthly payment plan over 4 years is offered for the car.

b) How much would be paid each month during this time?

3 marks

Total marks ……../18

Reasoning Test 14   Name _____

1 Circle the fraction that represents 0.4.

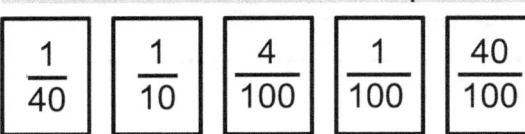

                                                                    1 mark

2 This tables shows the results of a javelin throwing competition.

Competitor	Attempt 1 (m)	Attempt 2 (m)	Attempt 3 (m)
Competitor A	65.5	66.2	64.9
Competitor B	68	67.2	66.8
Competitor C	70	68.6	67.4
Competitor D	65.9	66	67.4

a) Which competitor threw the javelin furthest?

b) What is the difference between Competitor D's shortest and longest throws?

m

                                                                    1 mark

Reasoning Test 14   Name _____

3 These shapes are regular hexagons.

Calculate the outer perimeter of this tessellated shape.

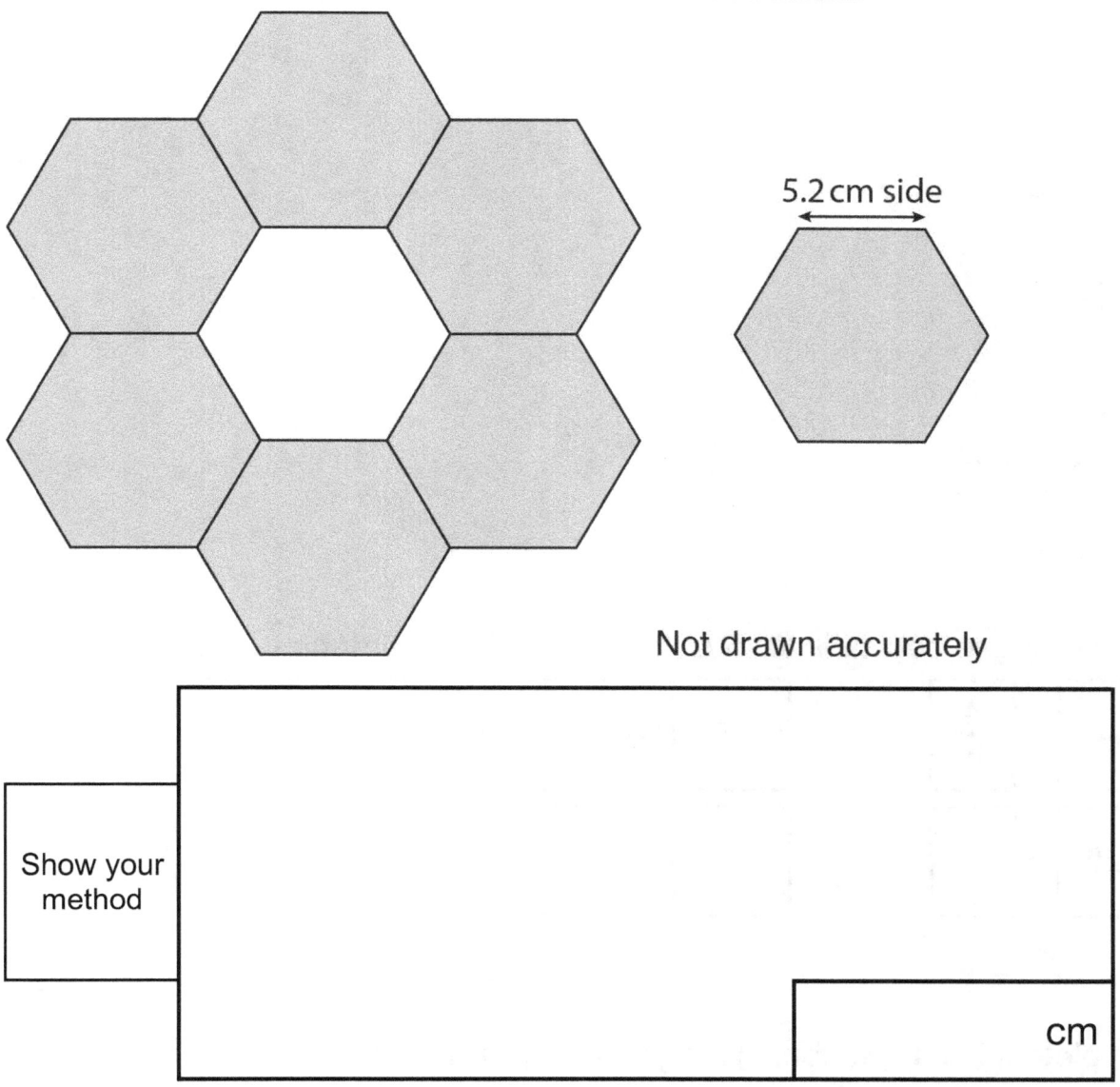

Not drawn accurately

Show your method

cm

1 mark

4 Plot these coordinates and join them in the correct order to make a polygon.

(0, 0) (2, 0) (1, 3) (3, 3) (0, 6) (2, 6)

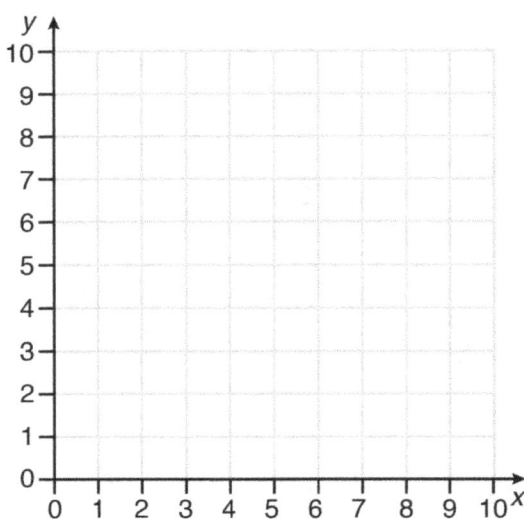

1 mark

5 Convert these decimals to fractions. One has been done for you.

$0.4 = \dfrac{4}{10}$ $0.75 = \boxed{}$ $0.3 = \boxed{}$

$0.25 = \boxed{}$ $0.125 = \boxed{}$

1 mark

6 Complete this calculation. Write your answer as a fraction.

$\dfrac{3}{2} - \dfrac{3}{4} = \boxed{}$

1 mark

7 Which is longer, 6 days or 140 hours?
Explain your answer.

1 mark

8 Josh and his five friends are going to the cinema to watch a film. They count their money and between them they have £50.

Cinema prices
Standard single ticket – £8.50
Group ticket (4 people) – £30

a) Can Josh and his friends afford to go to the cinema with the money they have?

Yes / No

Reasoning Test 14

A group ticket for 4 people costs £30.

b) How much money can they save by buying a group ticket rather than single tickets?

Show your method

£

2 marks

9 Calculate the value of z when x = 7.5 and y = 9.2.

(3x + 7) + (4y − 6) = z

Show your method

z =

2 marks

Reasoning Test 14   Name _____

10 Gabriella is making breakfast pots.

Breakfast pots
Ingredients (*makes 2 pots*)
165 g   yogurt
30 g   jam
10 g   sugar
15 g   granola
10 g   oats

a) How much yogurt does she need for 9 pots?

b) What is the ratio of granola to yogurt in each pot?

2 marks

Reasoning Test 14 Name _____

11 Mandy needs 1,200 m of wool for some scarves she is knitting.

a) If wool come in balls of 75 m, how many balls of wool does she need to buy?

balls

Wool costs £3.25 per ball with a 10% discount for orders of 10 or more balls.

b) If Mandy buys 20 balls of wool how much will she pay?

£

2 marks

12 A new mobile phone contract lasts for 24 months. Zara cannot decide whether to take a contract and pay £15 per month or to buy the phone outright for £335.

a) Which is the cheaper option for Zara?

Explain your answer.

Reasoning Test 14   Name _____

The contract includes 150 minutes of talk time per month.

b) How much would talk cost per minute if Zara chose the contract option?

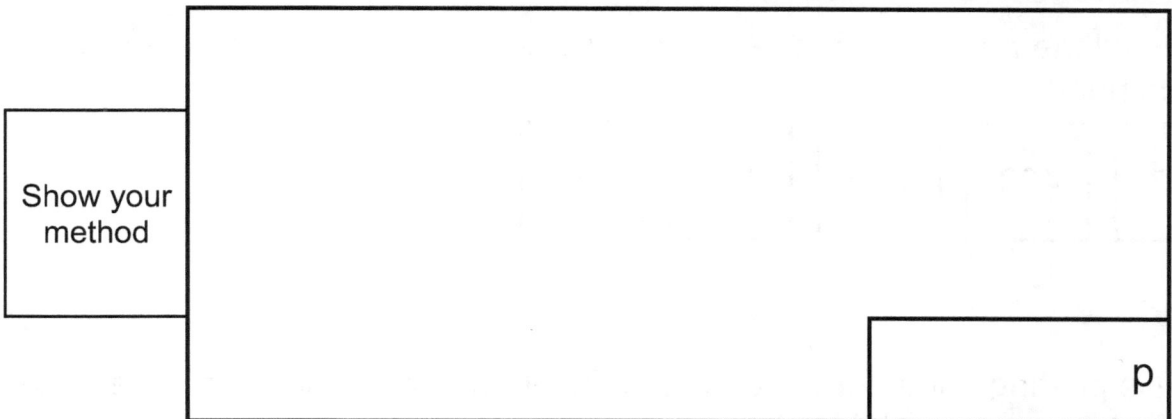

3 marks

Total marks ..../18

Reasoning Test 15

Name _____

1 Here is a number sequence.

9   18   27   36   45

Circle all the number cards below that would appear in the sequence if it continued.

| 117 | 400 | 500 | 450 | 99 |

1 mark

2 Luke is posting some letters to friends in different countries. He is sending two first class letters to Australia and one second class letter to Europe.

How much postage will he pay?

International postage charges – standard letter

Country	First class	Second class
Australia	£1.96	£1.37
South Africa	£2.50	£1.75
Europe	£1.30	£0.91
Canada and USA	£2.15	£1.50

Show your method

£ _____

1 mark

9 Find three consecutive numbers between 10 and 20 that add to make a square number.

Explain your answer.

2 marks

10 At the gym, James runs on a treadmill for 400 m then does 35 press-ups and 20 burpees.

He completes 12 rounds of this sequence.

a) Fill in the missing information in the table.

Round	1	12
Run	400 m	m
Press-ups	35	
Burpees	20	240

b) If it takes him an average of 4 minutes 30 seconds to complete one round, can he do all 12 rounds in 45 minutes?

Explain your reasoning.

Yes / No

2 marks

Reasoning Test 15   Name _____

11 A netball court is split into three equal parts. The length of the court is double its width.

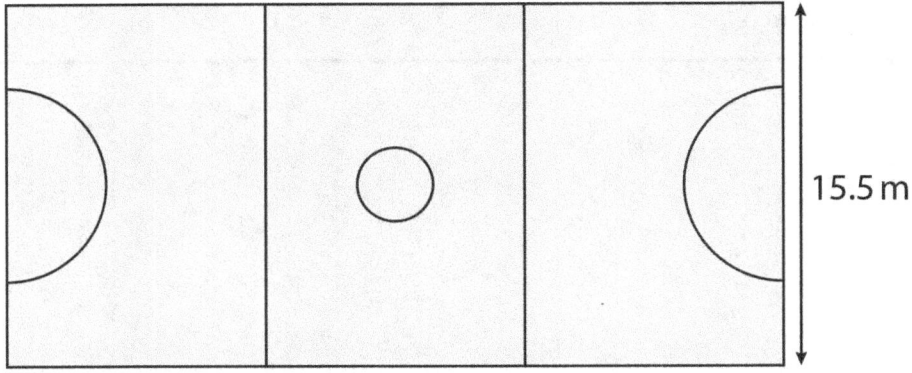

Not drawn accurately

a) What is the length of the netball court?

Show your method

m

b) What is the area of the netball court?

Show your method

m²

2 marks

12 An orchard grows apples to make apple juice. It takes 8 apples to make 250 ml of juice.

a) How many apples will be required to make 8 litres of juice?

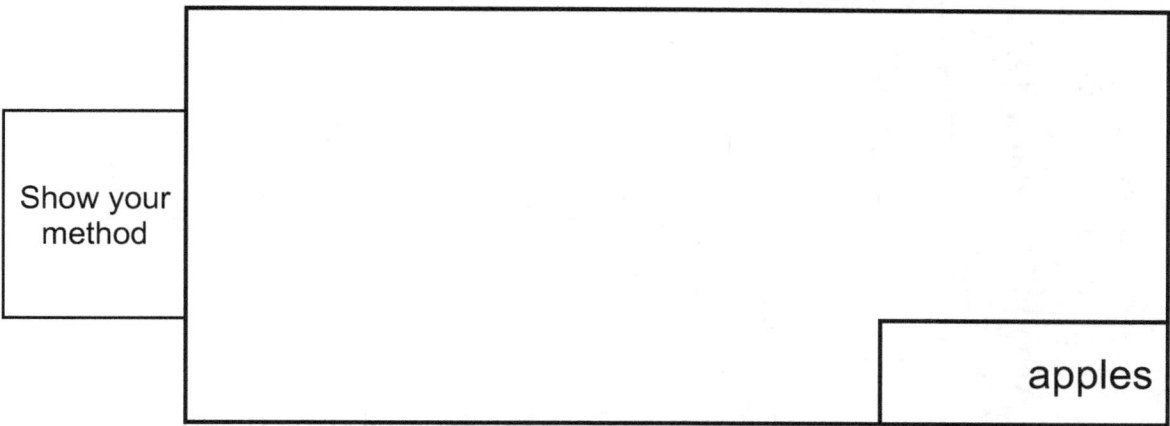

apples

The apple juice is sold in bottles holding 250 ml.

b) If each bottle costs 25p to make and is sold for £2.50, how much will profit will 50 bottles make?

£

3 marks

Total marks ..../18

Reasoning Test 16

Name _____

1 Circle the jugs that are $\frac{3}{4}$ full.

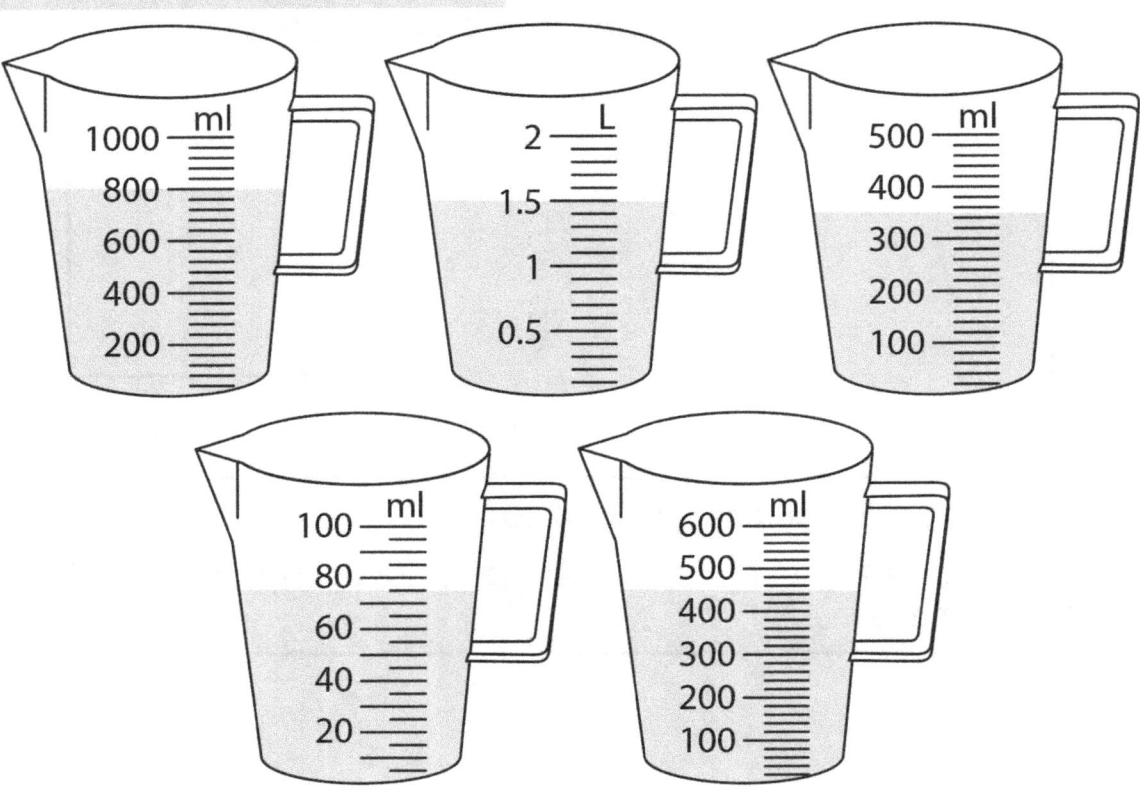

1 mark

2 This table shows the average temperatures in January in two different cities.

What is the difference in temperatures between Edinburgh and Cape Town at 2 a.m.?

January
Average temperature (°C)

Time	Edinburgh	Cape Town
12 a.m.	−8	13
2 a.m.	−9	14
4 a.m.	−8	16
6 a.m.	−7	18
8 a.m.	−6	21
10 a.m.	−3	23

Show your method

____ °C

1 mark

Reasoning Test 16   Name _____

3 Draw two parallel lines that are exactly 1 cm apart.

Line A must be 7.5 cm long and line B must be 4.5 cm long.

Use a ruler.

1 mark

4 Draw a quadrilateral with two acute angles and two obtuse angles.

Use a ruler.

1 mark

5 Complete the fractions equation.

$\frac{7}{14} + \boxed{} = 1\frac{1}{4}$

1 mark

6 Write the missing fraction to complete this equation.

$\frac{1}{2} + \frac{3}{4} = \frac{7}{8} + \boxed{}$

1 mark

Reasoning Test 16   Name _____

7 'The mean average of the following numbers is 50.'

82   4   108   46   15

Is this statement true or false?

Explain how you know.

Show your method

True / False

1 mark

8 Jack plants a number of seeds in a tray.

$\frac{7}{12}$ of the seeds are pansies.

A Pansies 63	B Lettuces 27
	C

a) What fraction of the seeds are lettuces?

Use a bar model to help answer this question.

Show your method

Reasoning Test 16   Name _____

He has space left in section C so he decides to plant tomato seeds.

b) How many seeds can he plant in section C?

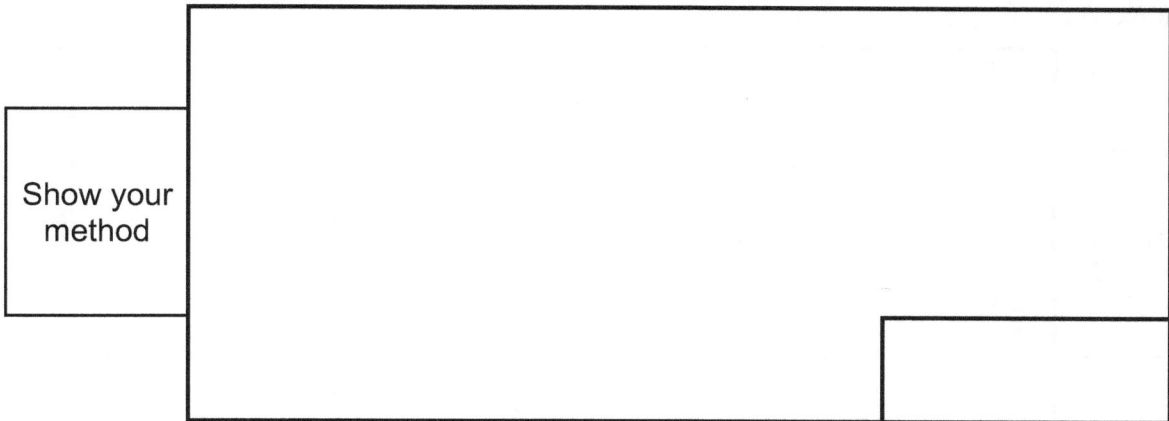

Show your method

_____
2 marks

9 In this grid, you multiply two numbers across the rows or down the columns to calculate the numbers in the grey boxes.

Use four numbers from 1 to 9 to complete the grid.

☐ ×	☐	8
☐	× ☐	27
6	36	

_____
2 marks

Reasoning Test 16   Name _____

10 3,550 people buy tickets to attend a concert.

If the tickets cost £5.50 each, how much money is collected in total for tickets?

£

2 marks

11 Triangle A is an enlargement of Triangle B. The scale factor is 3.5.

a) What is the missing base length of Triangle B?

Not drawn accurately

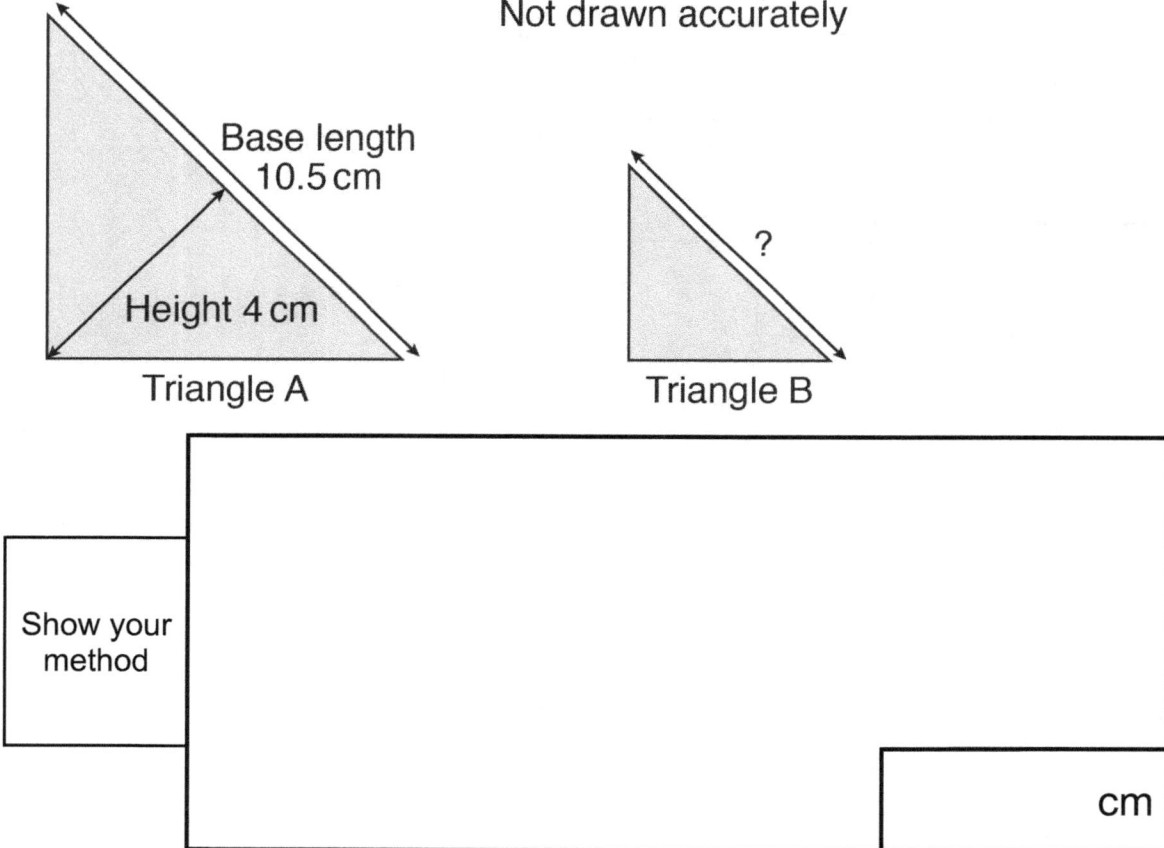

cm

b) What is the area of Triangle A?

cm²

2 marks

12 All 50 children in Year 6 go on a school trip to the theme park. They pay £5 each if they go on the coach or £3 each if their parents drop them off. In total, they pay £230.

Example:
4 children travelling on the coach pay £20
10 children travelling with parents pay £30
Total £50

a) How many children go on the coach and how many go with their parents?

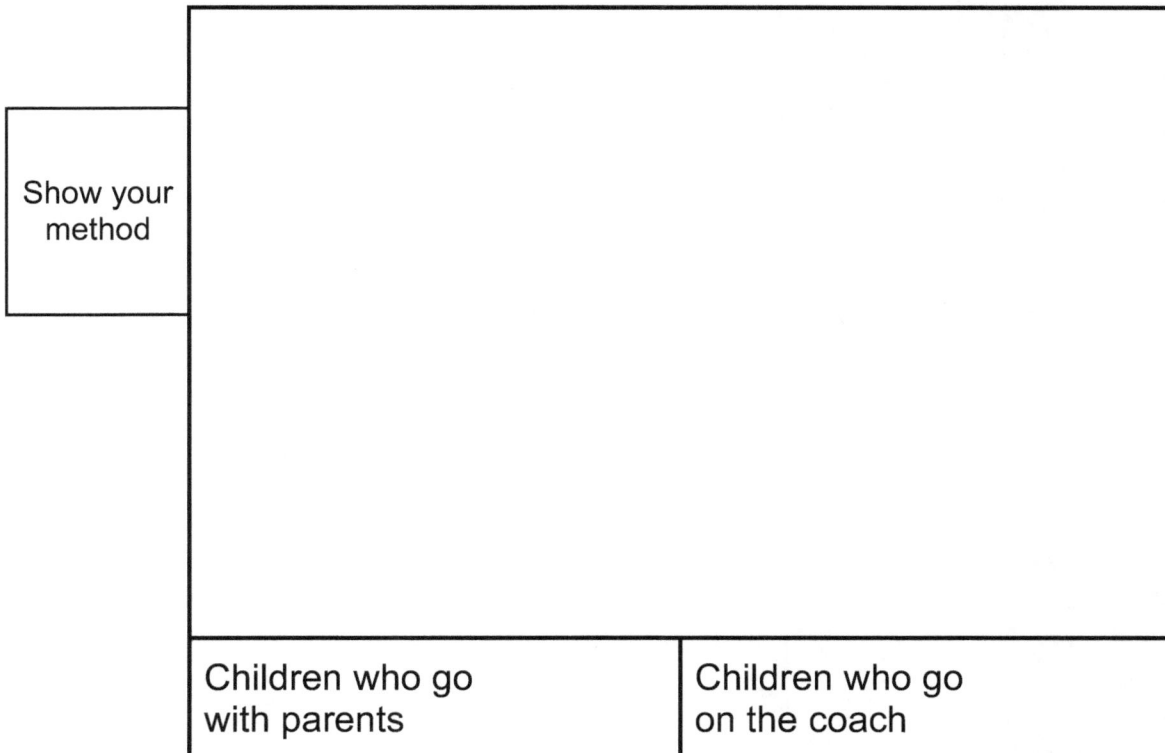

Children who go with parents | Children who go on the coach

Reasoning Test 16   Name _____

b) $\frac{3}{10}$ of the children on the trip each buy a never-ending slushy for £5.99.

How much do they spend in total on slushies?

Show your method

£

3 marks

Total marks ………/18

Reasoning Test 17   Name _____

1   Which percentage is equal to $\frac{4}{5}$?

    Circle the correct answer.

    20%   40%   65%   8%   80%   16%

                                                                        1 mark

2   The table shows the populations of five different countries.

Populations of countries	
UK	65,788,574
South Africa	56,015,473
Spain	46,461,567
Canada	36,289,822
Australia	24,125,848

    Round each population to the nearest million.

    Then use your rounded numbers to complete the bar chart.

    One bar has been done for you.

    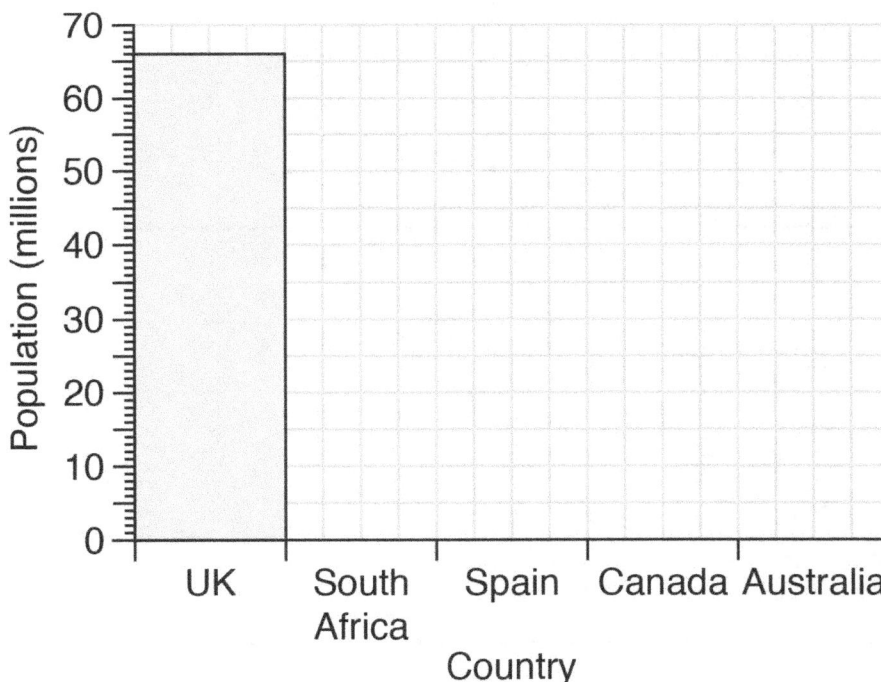

                                                                        1 mark

Reasoning Test 17 Name _____

3 Convert 289 minutes to hours and minutes.

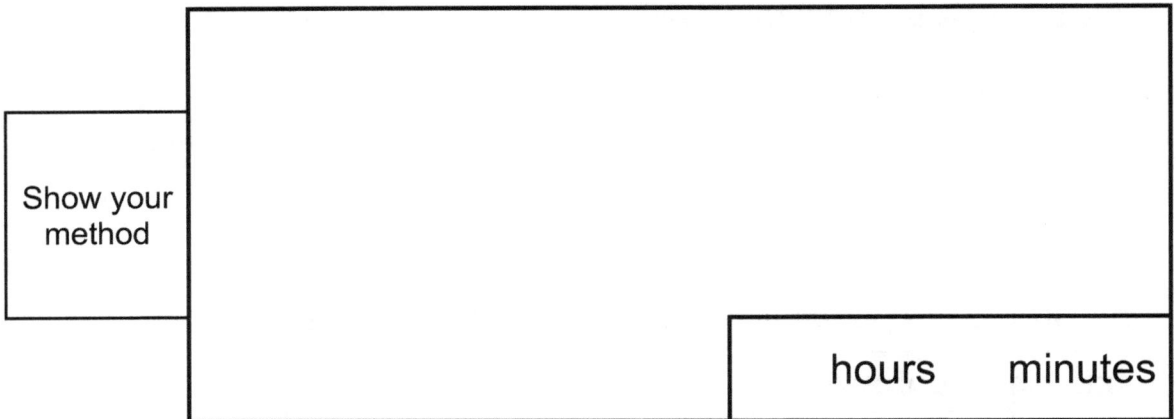

hours   minutes

1 mark

4 Calculate the missing angle in the pentagon.

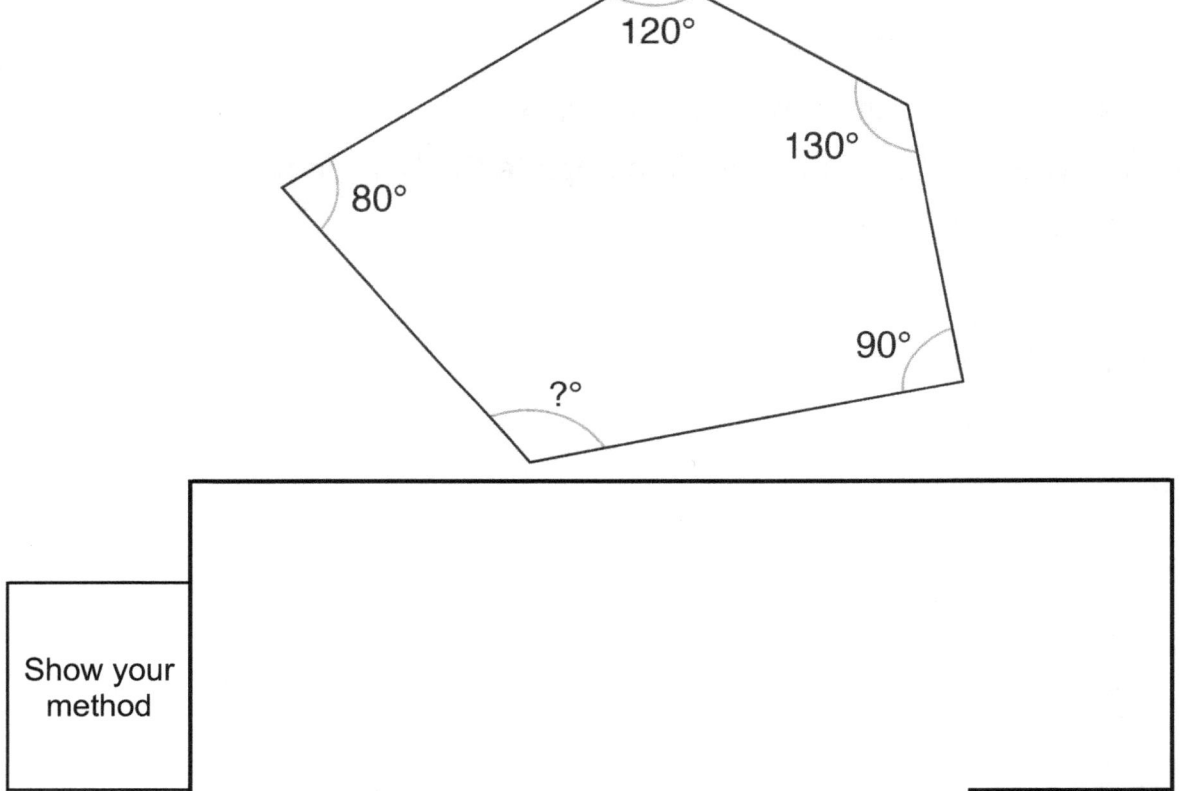

1 mark

Reasoning Test 17   Name _____

5 Complete this calculation.

$1\frac{1}{2} \div \frac{1}{4} =$

1 mark

6 Join each decimal fraction to its equivalent proper fraction.
One has been done for you.

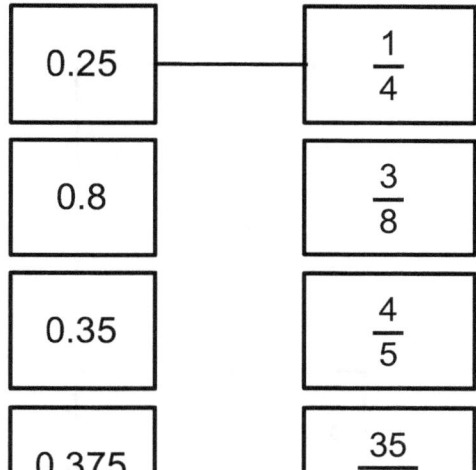

1 mark

Reasoning Test 17   Name _____

7 Which is greater, 840 seconds or 13 minutes?
Show your calculations.
Use a 'greater than' or 'less than' sign to write your answer.

1 mark

8 Leonie exchanges some money from pounds to dollars. For each pound, she gets $1.40.

a) How many dollars will she get for £25?

b) If Leonie receives $21, how many pounds did she exchange?

2 marks

Reasoning Test 17   Name _____

9 In Oak Class, 24 of the children are boys and a quarter of the children in the class are girls.

Altogether, how many pupils are there in the class?

You could use a bar model to support your calculations.

Show your method

2 marks

10 A rabbit eats 3 packs of carrots every 2 weeks. Each pack weighs 1.4 kg.

a) How many kilograms of carrots does the rabbit eat in a year?

Show your method

kg

Reasoning Test 17   Name _____

Each pack of carrots costs £1.15.

b) How much does it cost to feed the rabbit for 6 weeks?

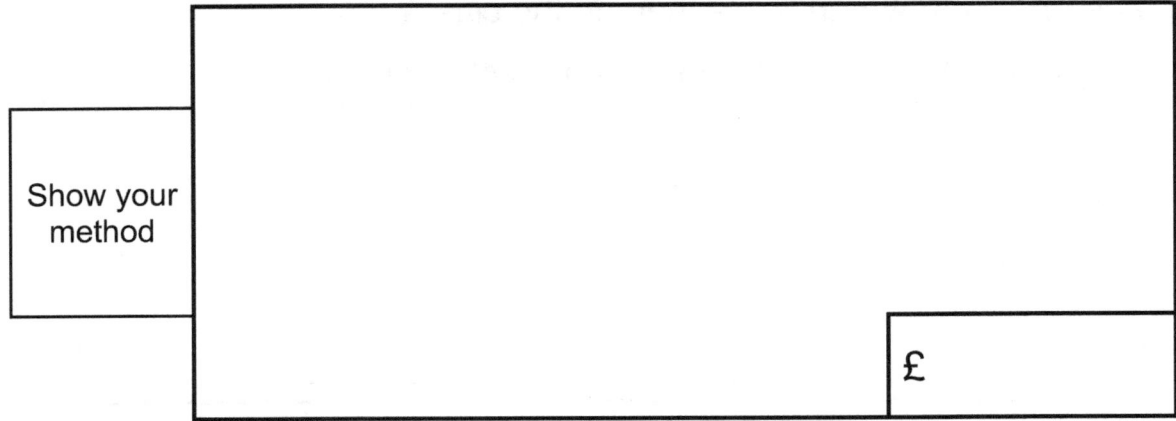

2 marks

11 Alex earns £3,200 per month. 20% of his earnings are taken in tax.

a) How much money does he have left after tax, per month?

b) Alex gets a 15% pay rise. How much does he earn per month now?

2 marks

Reasoning Test 17   Name _____

12 A train is travelling 930 km through France, from Paris to Nice.

8 km = 5 miles (approximately)

a) How many miles is this journey?

miles

b) The journey takes 9 hours. How fast is the train travelling in km per hour? Round your answer to the nearest whole number.

km/h

3 marks

Total marks ………/18

Reasoning Test 18

Name _____

1 Write the next three numbers in the sequence.

| 50 | 75 | 100 | 125 | | | |

1 mark

2 The table shows some exchange rates.

Spend	£400+	£500+	£1000+
Euro (€)	1.0843	1.0954	1.0966
US dollar (US$)	1.2435	1.2564	1.2576
UAE dirham	4.4759	4.4900	4.5018
Australian dollar (Aus$)	1.7186	1.7257	1.7275
New Zealand dollar (NZ$)	1.8081	1.8109	1.8137

If you changed £500 into Australian dollars, what would you receive for each pound? Round your answer to two decimal places.

$ _____

1 mark

3 This symmetrical shape was made by reflecting one side in the mirror line. Calculate its perimeter.

Use a ruler to check for matching lengths.

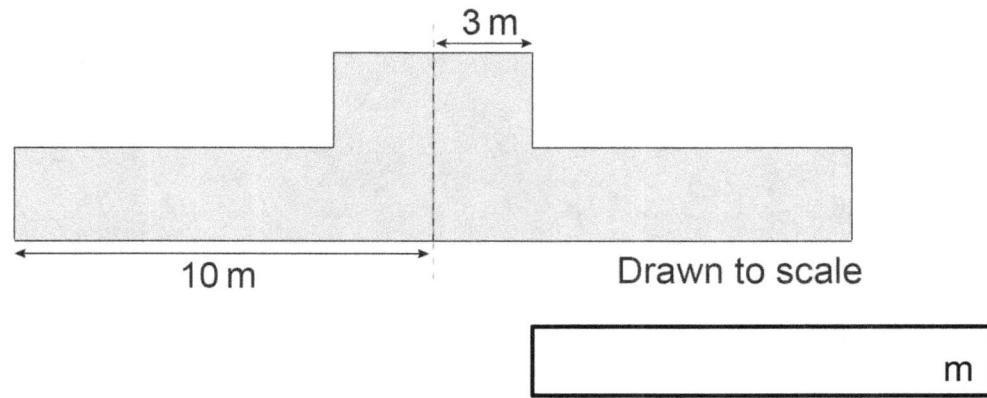

Drawn to scale

_____ m

1 mark

Reasoning Test 18   Name _____

4 Nancy leaves for work at 09:30. She returns 10 hours later.

On the clock, draw the time that Nancy returns.

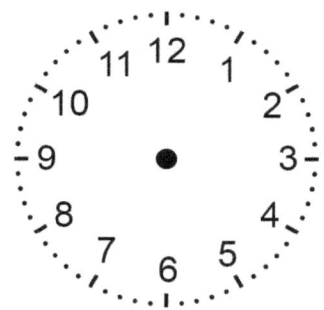

1 mark

5 $\frac{3}{10}$ of a number is 10.5. What is the original number?

1 mark

6 Convert these fractions to thirtieths.

One has been done for you.

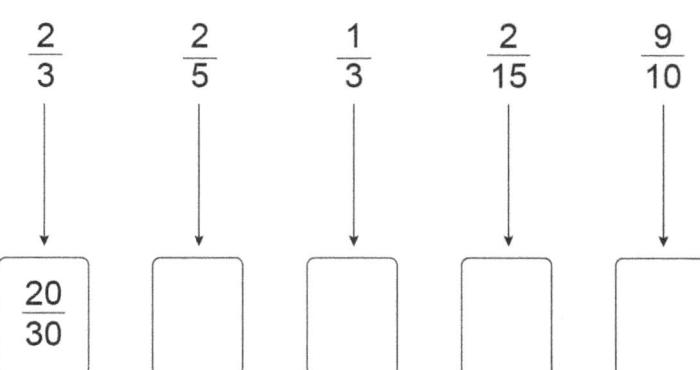

1 mark

7   The area of square A is 25% of the area of square B. If square A has a side of length 5 cm, what is the length of a side of square B?

Explain your answer.

cm

1 mark

8   A cube-shaped container has edges of 6 cm. Paul fills the cube with water, then pours the water into a cuboid-shaped container with a square base of 3 cm.

What height will the water reach in the cuboid container?

cm

2 marks

9 72 pens are divided equally into different numbers of pots. Sophie thinks there could be 10 different ways of dividing the pens, with more than one pen in each pot.

Explain why she is correct.

Show your method

2 marks

10 Tom has 1.2 m of ribbon. He cuts it into two pieces. One piece of ribbon is 30 cm longer than the other piece.

What is the length of each piece of ribbon?

You could use a bar model to work this out.

120 cm ribbon

Show your method

Piece 1 =   cm   Piece 2 =   cm

2 marks

Reasoning Test 18   Name _____

11 An Italian cruise ship travelled 620 km. How many miles did the cruise ship sail?

1 mile = 1.6 km (approximately)

miles

2 marks

12 A car is travelling from England to Scotland. The journey is 425 miles long.

a) If the journey takes 8.5 hours, what is the average speed of the car?

mph

Petrol costs £1.34 per litre. On this kind of journey, one litre will be used every 5 miles.

b) How much does the fuel for the car on this journey cost?

Show your method

£

3 marks

Total marks ..../18

Reasoning Test 19   Name _____

1. Place these fractions on the number line, from smallest to largest.

$$\frac{2}{3} \quad \frac{1}{3} \quad \frac{1}{2} \quad \frac{6}{6} \quad \frac{5}{6}$$

0 ———————————————————————— 1

1 mark

2. This pictogram shows the number of cars sold by a garage throughout one year, month by month.

Month of the year	Cars sold in month
January	2½
February	3½
March	4½
April	4½
May	5
June	2½
July	1½
August	1
September	2½
October	3
November	2½
December	1½

🚗 = 100 cars

What was the average number of cars sold per month, from January to April?

Show your method

_____ cars

1 mark

Reasoning Test 19   Name _____

3 What shape has this net?
Circle the correct answer.

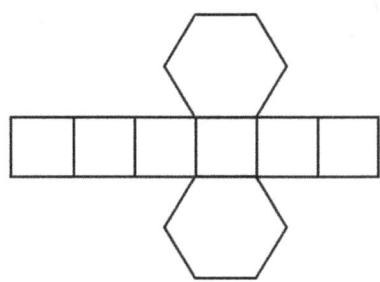

| Pentagonal pyramid | Square-based pyramid | Octagonal prism | Hexahedron | Hexagonal prism |

1 mark

4 Rotate this shape 180° around its centre point. Draw the shape in its new position.

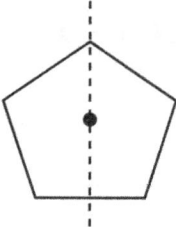

Regular pentagon

1 mark

5 Complete this calculation.

$(\frac{4}{9} \times \frac{3}{4}) + (\frac{1}{3} \times \frac{2}{3}) = \square$

1 mark

6 Change this improper fraction into a mixed number.

$\frac{131}{4} = \square$

1 mark

Reasoning Test 19   Name _____

7 Amber is thinking of a secret number. The product of her secret number and 30 is 2,250.

What is Amber's secret number?

Explain your answer.

1 mark

8 A loaf of bread has 16 slices.

a) If the price of the loaf is £1.44, how much does each slice of bread cost?

pence

b) If the price of each slice increases to 12.5p, what is the new total price of the loaf?

£

2 marks

9 Find the value of the letter in each calculation.

a) $a - 86.4 = 2.8$

b) $0.27 + b = 5$

2 marks

10 Lucy bought a sculpture for £3,500. She sold it 3 months later at a profit of 35%.

a) How much did she sell the sculpture for?

b) If Lucy had sold the sculpture for £2,800, by what percentage would its value have decreased?

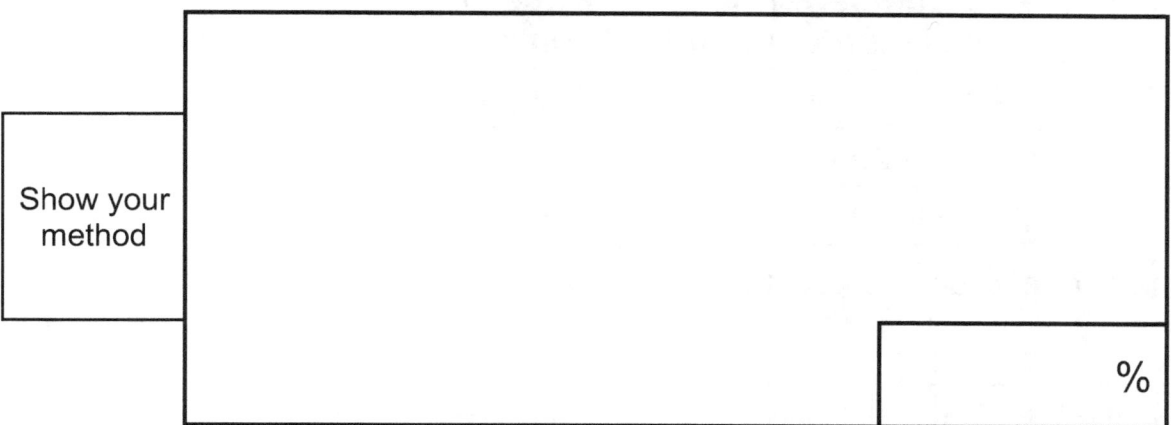

2 marks

11. The length of one side of a rectangle is 48 cm. The length of its adjacent side is $\frac{3}{4}$ of this.

a) How long is the adjacent side?

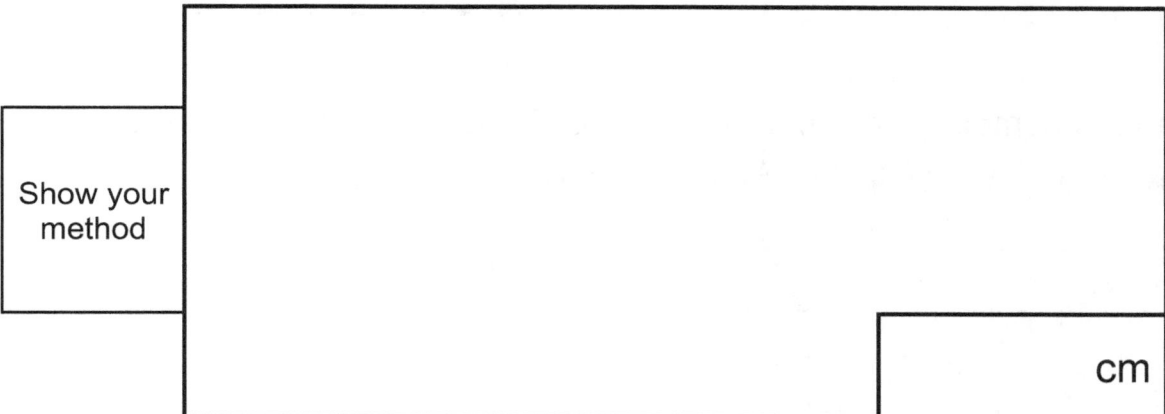

b) What is the area of the rectangle?

2 marks

12 A school orders some new computers.

Number of computers bought	Price per computer (£)
< 10	419.00
10–20	389.00
21–50	349.00

a) The school orders 23 computers. How much do they pay in total?

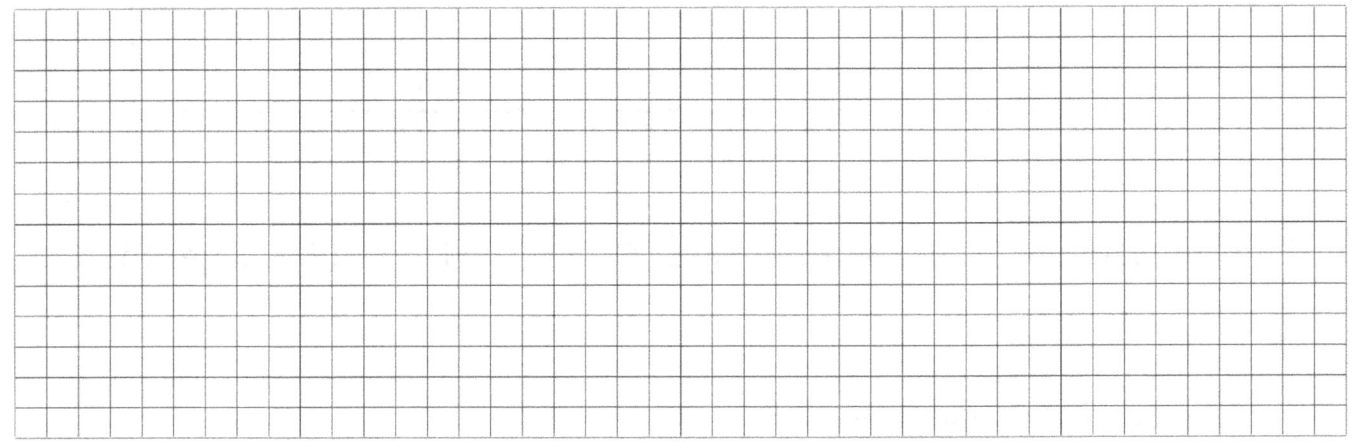

£ _____

When the company sells computers at £389.00, 45% of this price is profit.

b) How much profit will they make on 10 of these computers?

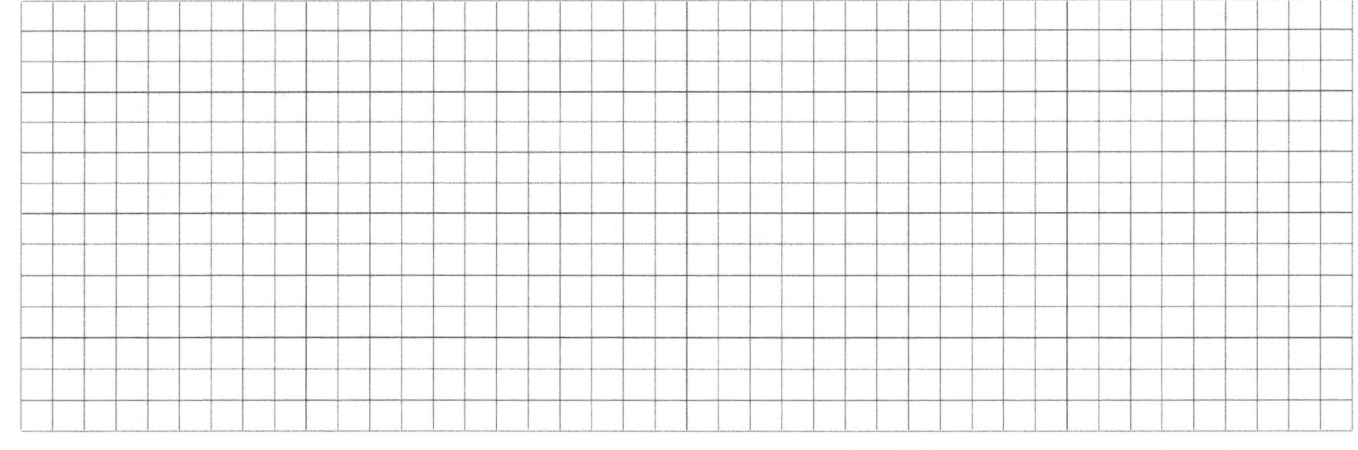

£ _____

3 marks

Total marks ………/18

Reasoning Test 20

Name _____

1 Order these lengths, from smallest to largest.

| 13 mm | 1.8 m | 12 cm | 1.2 m | 25 mm |

Smallest ☐ ☐ ☐ ☐ ☐ Largest

1 mark

2 This table shows the numbers of medals won and distances jumped by long jump competitors.

Year	Gold	Silver	Bronze	USA medals
1983	8.55 m (USA)	8.29 m (USA)	8.12 m (USA)	3
1987	8.67 m (USA)	8.53 m (URS)	8.33 m (USA)	2
1991	8.95 m (USA)	8.91 m (USA)	8.42 m (USA)	3
1993	8.59 m (USA)	8.16 m (RUS)	8.15 m (UKR)	1
1995	8.70 m (CUB)	8.30 m (JAM)	8.29 m (USA)	1

What is the mean average of the distances jumped in 1993?

Show your method

_____ m

1 mark

Reasoning Test 20

Name _____

3 Measure this parallelogram and calculate its area.

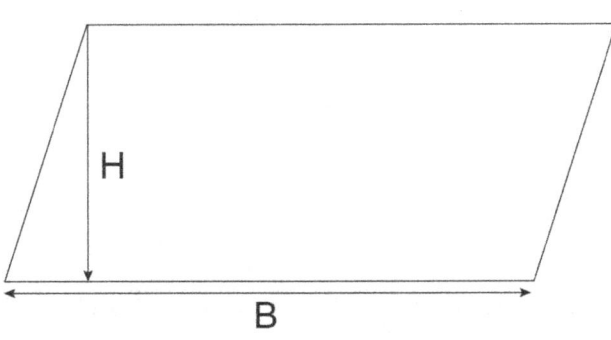

Area = B × H

Show your method

cm²

1 mark

4 Calculate the volume of this cube.

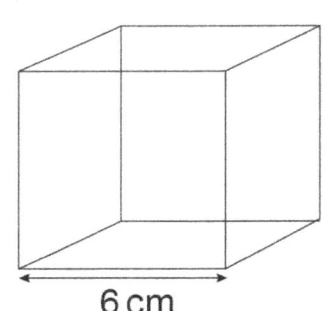

6 cm

Show your method

cm³

1 mark

Reasoning Test 20   Name _____

5 What is 3 divided by 12?
Circle the correct answer.

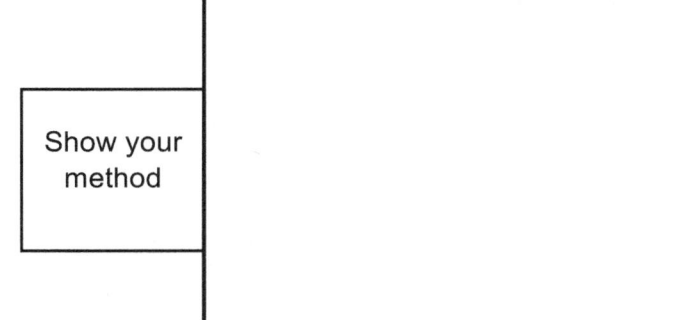

4   3   0.75   0.36   0.25

1 mark

6 Show whether each equation is true or false.

a) $\dfrac{7}{4} + \dfrac{3}{4} < 2$

Show your method

True / False

b) $\dfrac{2}{9} < \dfrac{1}{9} + \dfrac{4}{9}$

Show your method

True / False

c) $\dfrac{5}{8} = \dfrac{8}{40}$

Show your method

True / False

1 mark

Reasoning Test 20 Name _____

7 Heidi and Lewis are looking at this shape.
   Heidi says it's a rectangle, Lewis says it is not.

Who is right, Heidi or Lewis?
Explain your answer.

Show your method

m

1 mark

8 Ava has $\frac{3}{4}$ of a litre of orange juice.

a) If she shared the orange juice equally among 4 cups, how many millilitres would there be in each cup?

Show your method

ml

b) If Ava had 6 cups, each with a capacity of 250 ml, how much more orange juice would she need to fill them all?

ml

2 marks

9 Sam picked some apples from her tree. She ate one and gave half of what was left to her neighbour. The next day she gave half of what she had to the other neighbour. The 4 remaining apples were fed to the horses.

How many apples did Sam pick from the tree?

apples

2 marks

Reasoning Test 20

Name _____

10 Calculate the surface area of this cuboid.

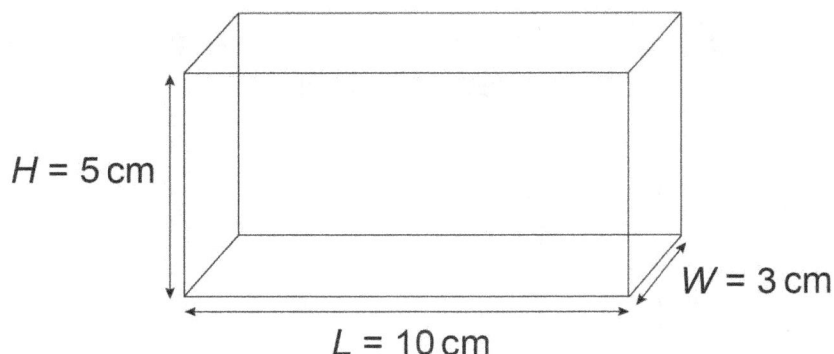

Show your method

_____ cm²

2 marks

11 a) How many cubes are there in this shape?

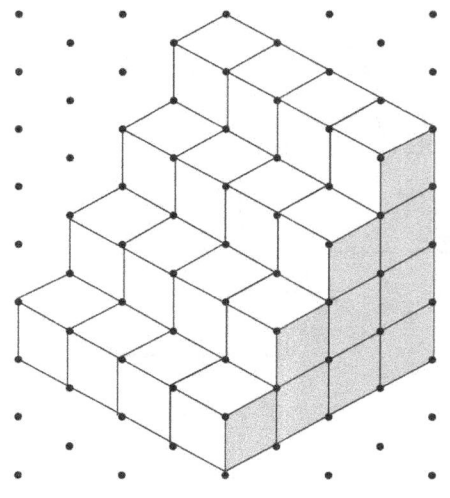

Show your method

_____ cubes

1 mark

Reasoning Test 20

Name _____

b) If the number of cubes in this structure was increased by a factor of 3.5, how many cubes would there be in the new shape?

cubes

1 mark

12 A family buys a new house for £255,000. They have to pay 5% tax on this amount.

a) How much tax do they pay?

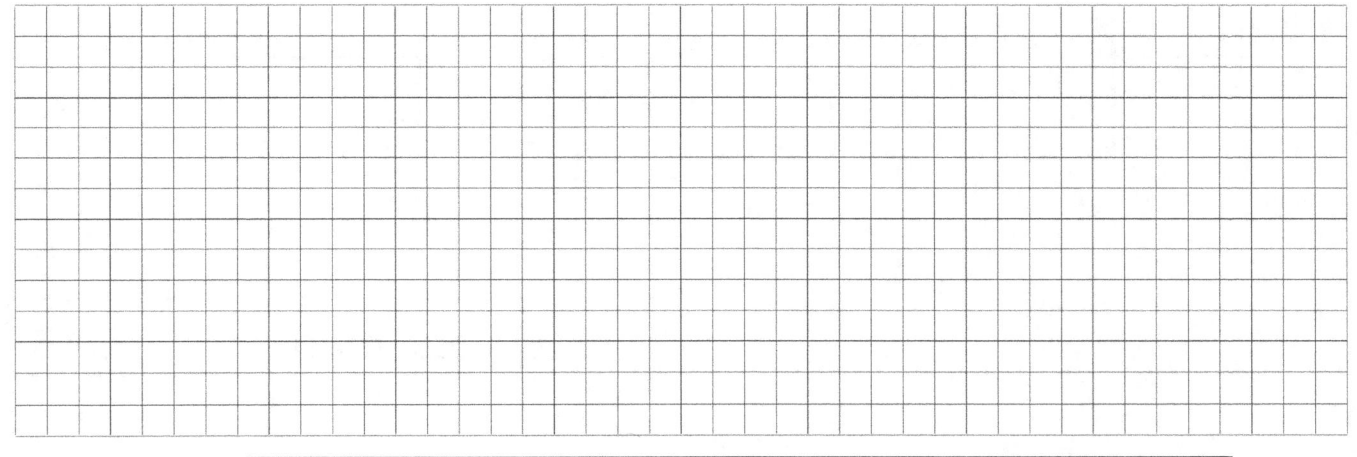

£

Reasoning Test 20

Name _____

They pay £935 each month for 25 years, to repay the loan for their house.

b) How much do they pay back in total?

Show your method

£

3 marks

Total marks ………/18

Reasoning Test 21   Name _____

1   Join the equivalent values in the table.
    One has been done for you.

0.7	$\frac{7}{1{,}000}$
0.007	$\frac{7}{10}$
0.7	$\frac{7}{100}$
7	$\frac{70}{100}$
0.07	$\frac{70}{10}$

1 mark

2   Julian has coloured in the dates he is going fishing in April.

```
 M    T    W    T    F    S    S
            1    2    3    4    5    6
 7    8    9   10   11   12   13
14   15   16   17   18   19   20
21   22   23   24   25   26   27
28   29   30
```

What fraction of the days in the month will Julian be fishing?

$\frac{\Box}{\Box}$

1 mark

3 What is the volume of this cube?

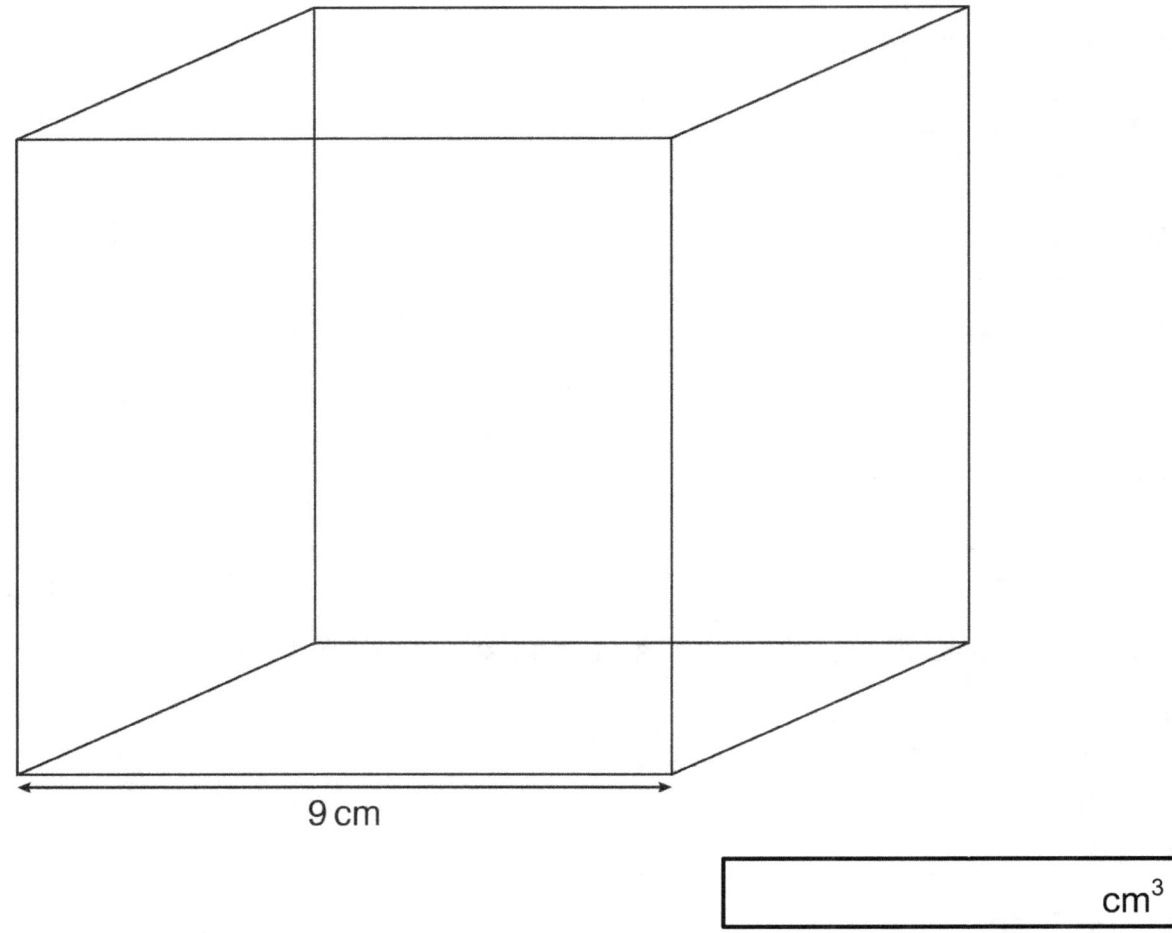

☐ cm³

1 mark

4 Complete this shape to make an octagon.

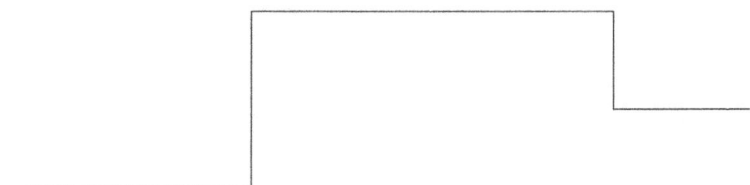

1 mark

Reasoning Test 21 Name _____

5 $\frac{2}{3}$ of a quantity is 28 litres. What is the whole quantity?

1 mark

6 Here is a fraction. In each empty box, write a different fraction that is equivalent to the one below.

1 mark

7 David switches on the washing machine on an economy cycle for 2 hours 42 minutes at 3.35 p.m.

He tells his wife it will be finished before he collects her from work at 6.10 p.m. Is he right?

Explain your answer.

1 mark

Reasoning Test 21

Name _____

8 Marcus ran a half marathon in 1 hour 44 minutes.

Half marathon = 13 miles approximately

How long did it take him to run each mile?

Show your method

_____ minutes

2 marks

9 Sarah and Imogen are given 48 centimetre cubes to arrange as rectangular shapes.

Work out the different perimeters can they make, using all the cubes?

Show your method

2 marks

Reasoning Test 21   Name _____

10 It takes 5 hours and 40 minutes to travel on the train from Cornwall to London.

  a) Max and Margaret arrive in London at 4.30 p.m. What time did their train leave Cornwall?

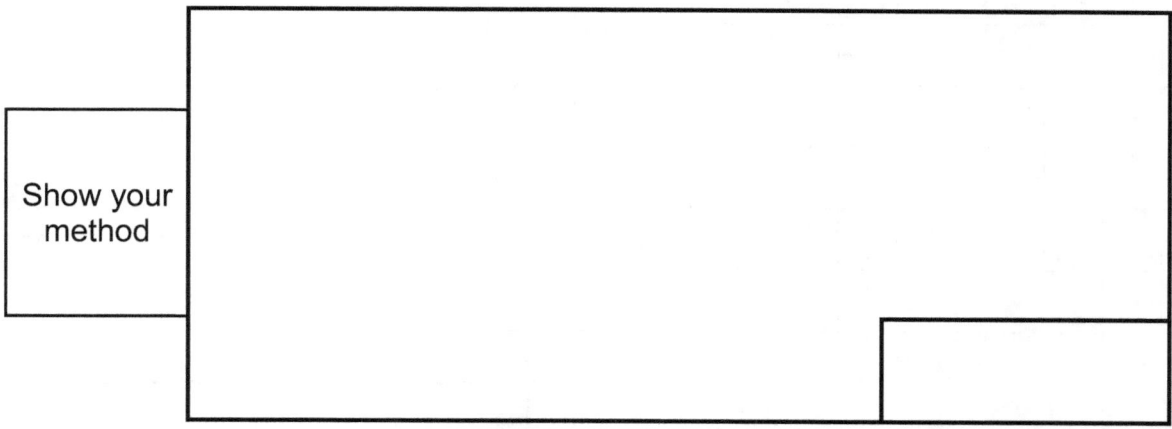

  b) They fly back from London and it is 4 times faster than the train. How long does the flight take?

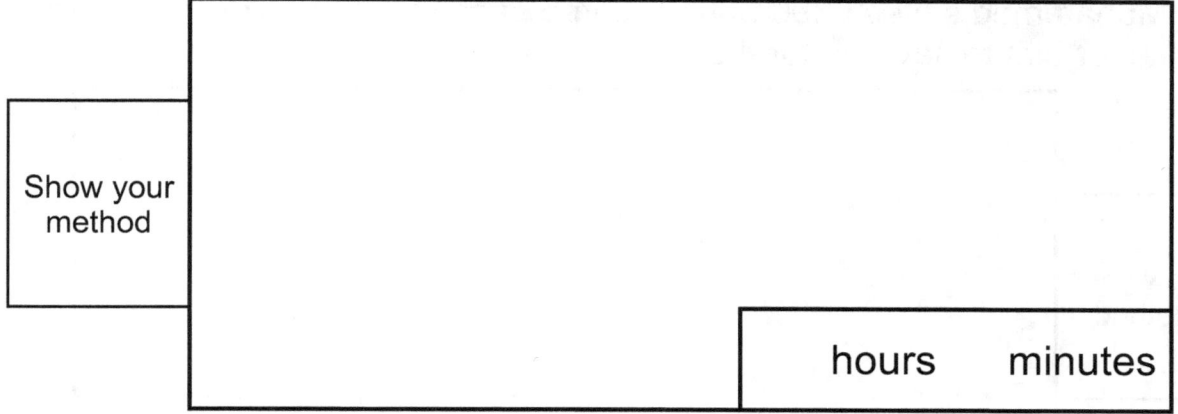

hours   minutes

2 marks

11 There are 60 marbles in a pot. Each marble is blue, red or green.

The ratio of colours (blue : red : green) is 3 : 4 : 5.

How many marbles of each colour are there in the pot?

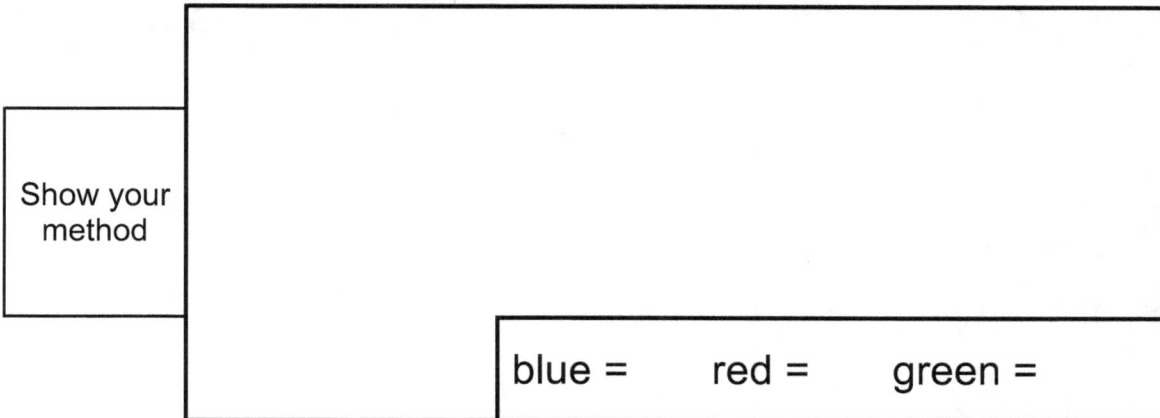

blue =   red =   green =

2 marks

Reasoning Test 21   Name _____

12 Mohammed works 37.5 hours per week and earns £26 per hour.

a) How much does Mohammed earn a week?

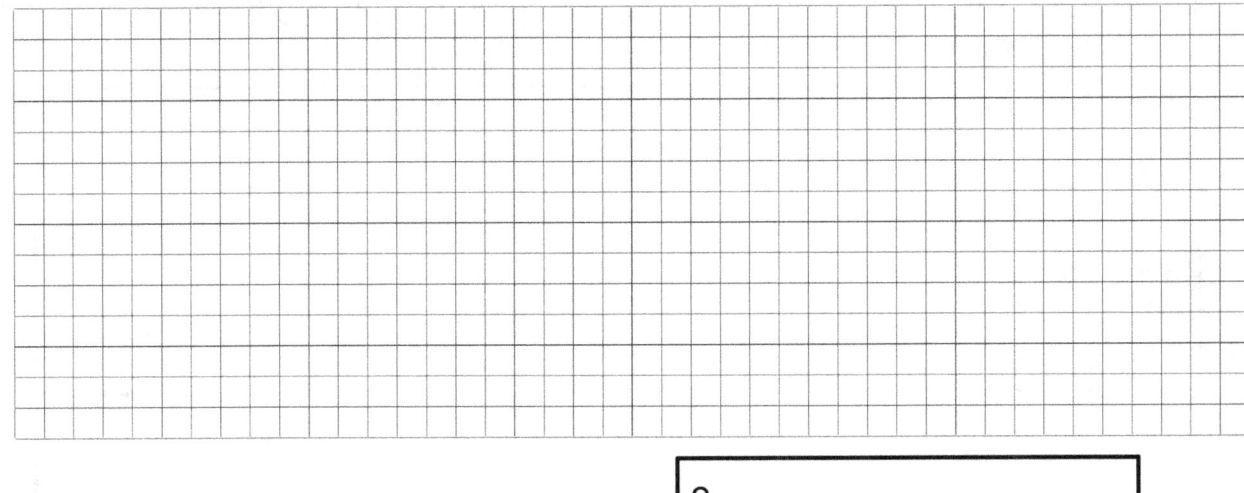

£ _____

He is saving for a new car that costs £5,000.

b) If Mohammed saves £390 of his income each week, how many weeks will it take for him to save up for the car?

Show your method

weeks

3 marks

Total marks ………/18

Reasoning Test 22

Name _____

1 Continue the pattern and fill in the number boxes.
One has been done for you.

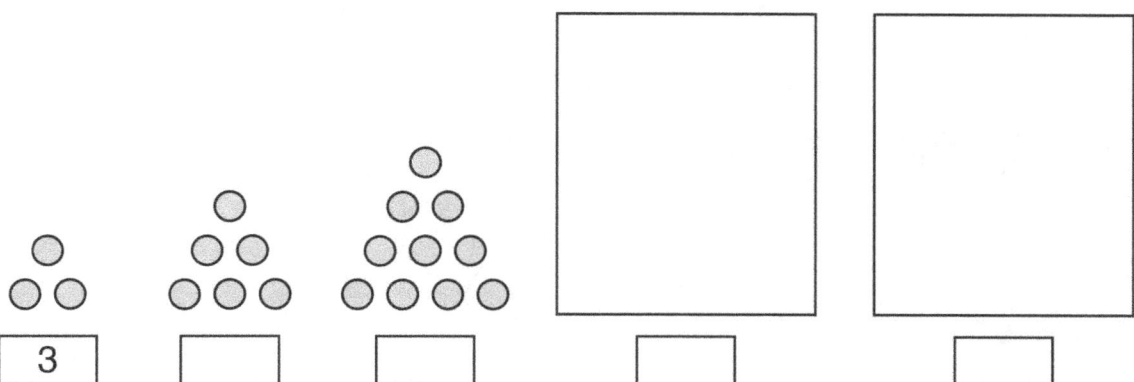

1 mark

2 This pie chart shows the numbers of points gained by football teams halfway through the season.

League points of football teams in top half of league table

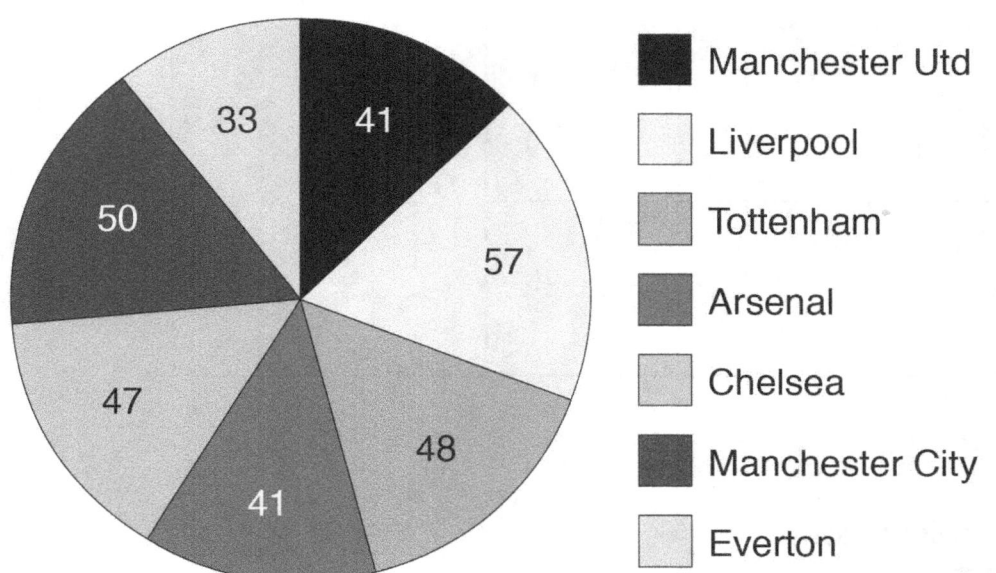

a) Which team is currently at the top of the league?

b) What is the difference in points between the highest and lowest teams in the top half of the league table?

_____ points

1 mark

Reasoning Test 22   Name _____

3 Draw the lines of symmetry on this shape.

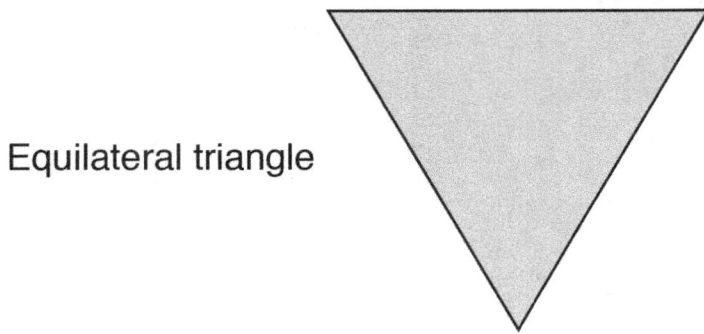

Equilateral triangle

1 mark

4 Does this shape contain more than 15 quadrilaterals?

1 mark

5 Complete this calculation.

$70 \times \dfrac{4}{5} = \boxed{}$

1 mark

Reasoning Test 22   Name _____

6 Circle all the fractions that are < 1.

$\boxed{\dfrac{3}{9}}$ $\boxed{\dfrac{3}{4}}$ $\boxed{\dfrac{12}{5}}$ $\boxed{\dfrac{89}{100}}$ $\boxed{\dfrac{16}{12}}$

1 mark

7 These two shapes are both quadrilaterals.

How do you know they are not the same shape?

Explain your answer.

Show your method

1 mark

Reasoning Test 22   Name _____

8. Dev is paving part of his back garden. He has chosen square slabs with side length 50 cm.

   How many slabs does he need to pave this part of his garden?

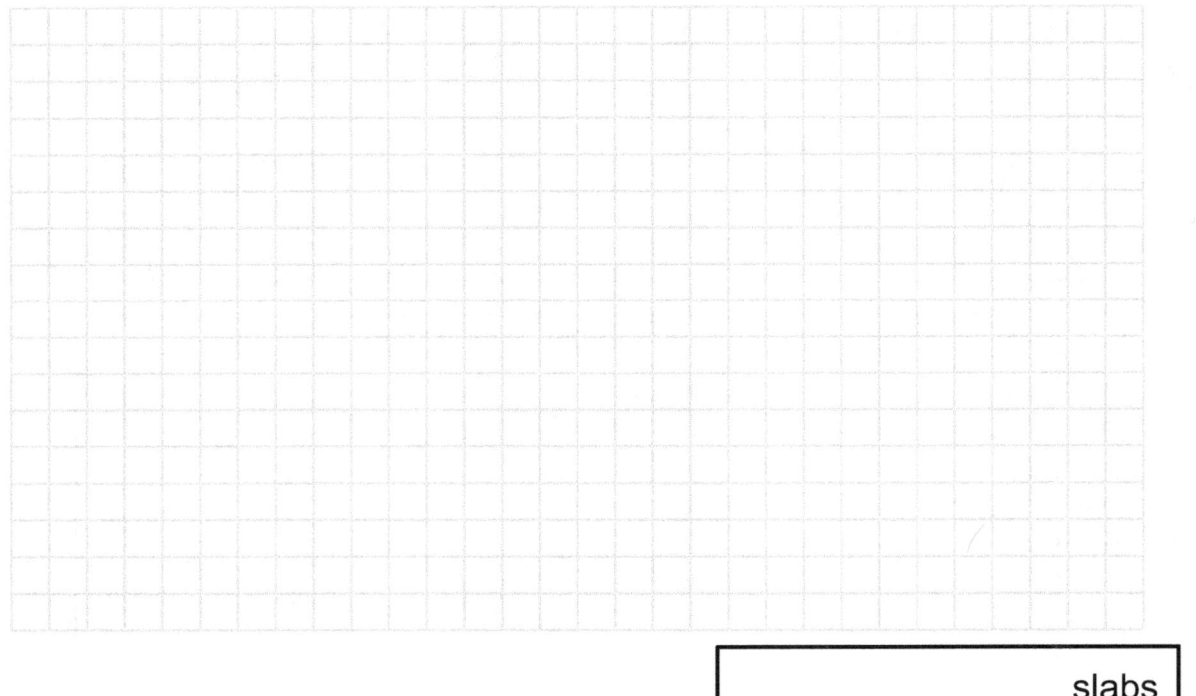

   [_____] slabs

   2 marks

9. Laura thinks of two numbers, she adds them together and divides the result by 2. Her answer is 46.

   One of her numbers is 24, what is the other number?

   Show your method

   2 marks

Reasoning Test 22   Name _____

10 a) Calculate the arrival time of a plane if it takes off at 21:18 and the flight lasts 6 hours 30 minutes.

b) If the plane travels at an average speed of 500 mph, how far does it fly?

miles

2 marks

11 Meg buys some milk at the supermarket. Her mum tells her that she needs 4 pints.

568 ml = 1 pint

The supermarket sells milk in one-litre boxes.

How many litres does Meg need to buy?

litres

2 marks

Reasoning Test 22   Name _____

12 Leanne makes wedding bouquets.

She buys 12 stems for each bouquet. Each stem costs 36p. The foliage is an extra £1.50 per bouquet.

a) How much does it cost Leanne to make one bouquet?

Show your method

£ _____

Leanne sells the bouquets for £25.99

b) How much profit does she make from selling 80 bouquets?

Show your method

£ _____

3 marks

Total marks ………/18

Reasoning Test 23 Name _____

1 Circle the numbers that are divisible by 3, 4 and 8.

| 102 | 96 | 48 | 124 | 24 |

1 mark

2 This chart shows how many times a dice landed on 1, 2, 3, 4, 5, 6 when it was thrown.

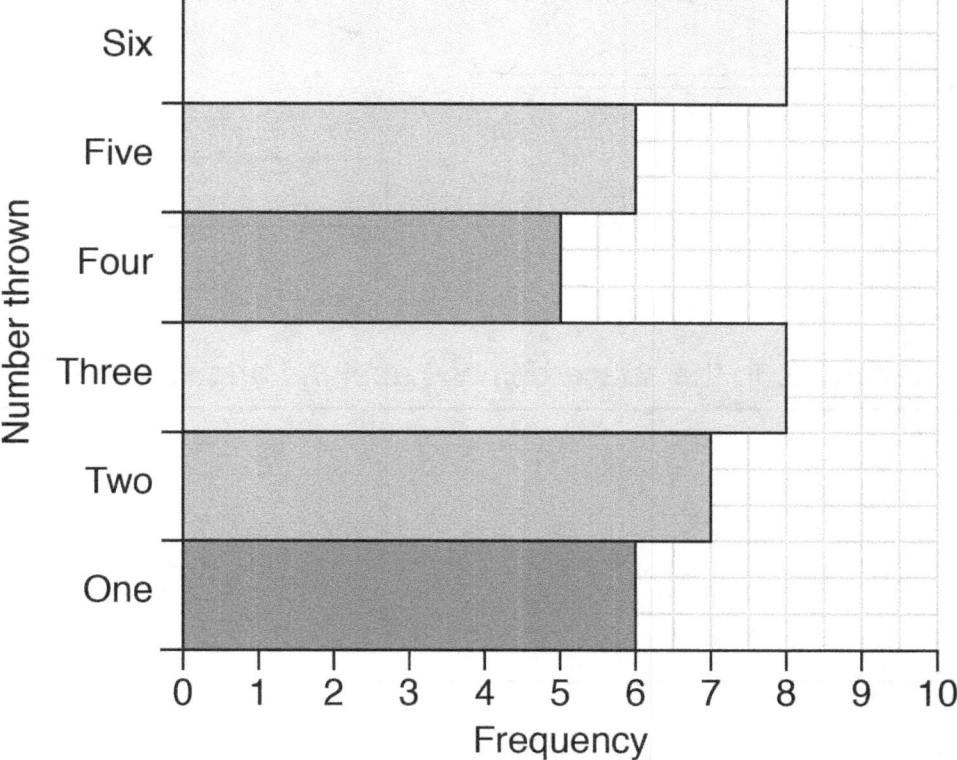

a) How many throws were made in total?

b) What was the range of the data?

1 mark

3 Calculate the total area of the two triangular faces of this shape.

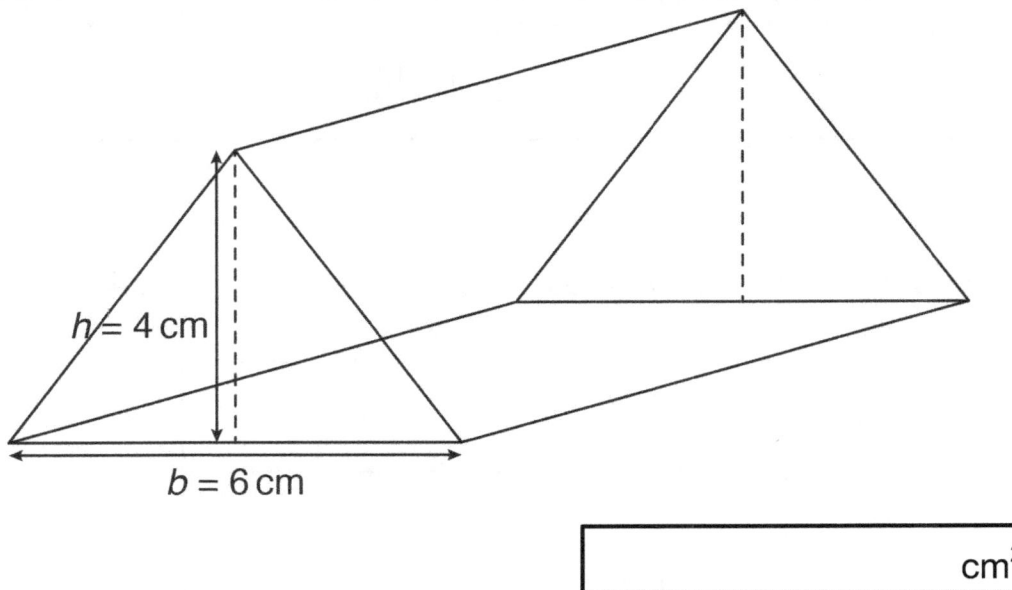

[ ] cm²

1 mark

4 Draw a reflex angle in the space below. Label the angle.

1 mark

5 Use common factors to calculate this fraction addition. Simplify your answer to its lowest terms.

$\frac{1}{3} + \frac{2}{5} =$

Show your method

1 mark

6 What is 60% of 580?

Show your method

1 mark

7 Marta adds three odd numbers. She says her answer is 40.

Why is Marta not correct?

Explain your answer.

Show your method

1 mark

Reasoning Test 23   Name _____

8 An academic diary comes in two sizes, type A and type B.
The type B diary has an area of 294 cm².
If the type A diary is 1.5 times the area of the type B diary, what is its area?

Show your method

2 marks

Reasoning Test 23

9 Jude makes a sequence of 5 numbers.

The first number is 3 and the last number is 31. He adds the same number each time.

Fill in the missing numbers.

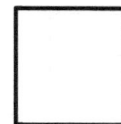 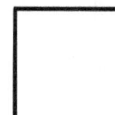 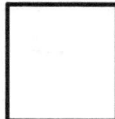

2 marks

10 The 420 children at a school go for lunch at different times.

$\frac{2}{7}$ of the children go at 11.45 a.m.

$\frac{2}{7}$ of the children go at 12.15 p.m.

The rest of the children go at 12.45 p.m.

How many children go for lunch at 12.45 p.m.?

Show your method

2 marks

Reasoning Test 23   Name _____

11 Which shape has the greater volume, shape A or B?

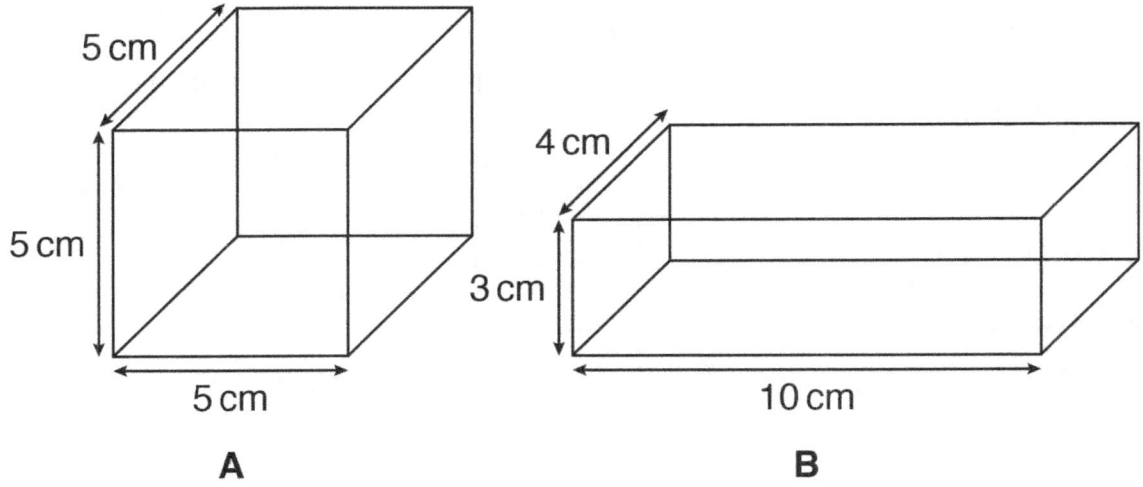

Circle the correct answer.

Explain how you know.

Show your method

A / B

2 marks

12 Mum is filling the bath for her 2 children. She tells them the bath will be ready by 6.45 p.m.

She turns on the taps at 6.30 p.m. and water flows into the bath at 6 litres per minute.

The bath takes 80 litres of water.

a) Will the bath be ready by 6.45 p.m.?

Explain your reasoning.

Show your method

Yes / No

b) If water costs 0.108p per litre, how much does it cost to have a bath?
Give your answer to two decimal places.

Show your method

p

3 marks

Total marks ..../18

Reasoning Test 24   Name _____

1 Rotate this shape 90° clockwise about point A.

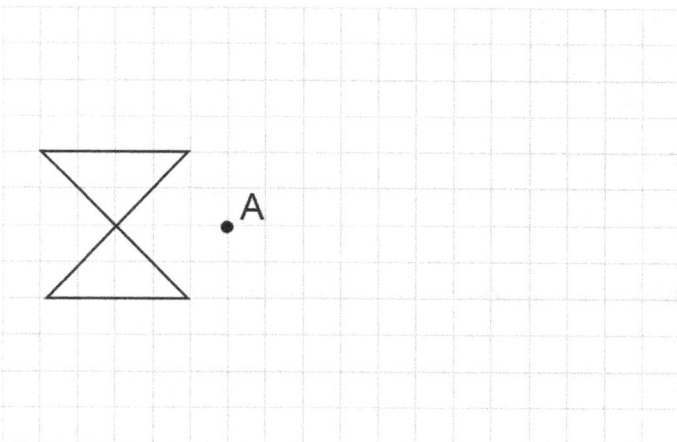

1 mark

2 This table shows the times that children in KS2 go to bed.

Time	Number of children
19:00–19:30	26
19:30–20:00	19
20:00–20:30	24
20:30–21:00	30
21:00+	21

a) How many children go to bed between 7.00 p.m. and 8.00 p.m.?

b) What fraction of the children go to bed between 8.30 p.m. and 9 p.m.?

1 mark

3 Label the three parts of this circle.

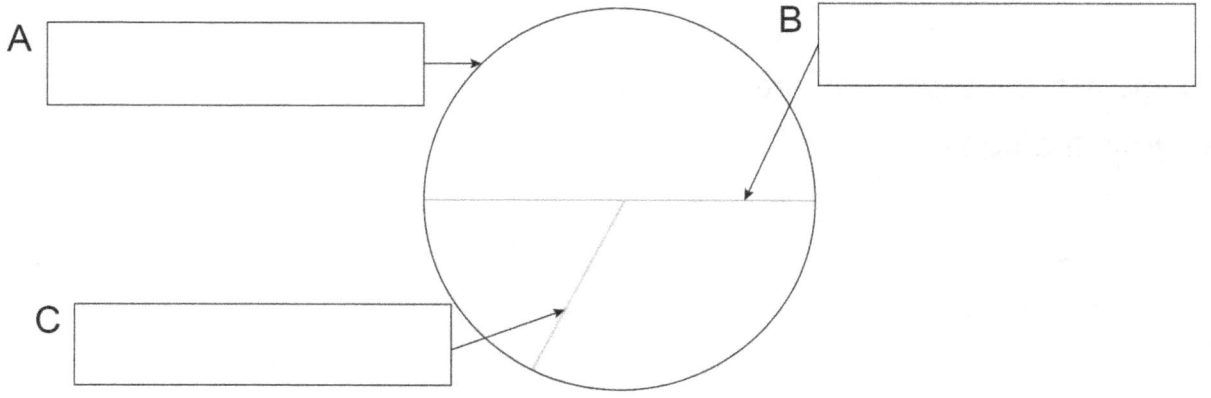

1 mark

Reasoning Test 24 Name _____

4 Toby describes a shape to his friend – he says:

> It has four sides of equal length, four angles, two sets of parallel lines, two lines of symmetry.

What is the shape that Toby is describing?

1 mark

5 **What fraction of 72 is 60?**
Give your answer in its lowest terms.

Show your method

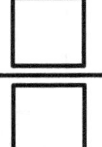

1 mark

6 **How many hundredths are there in 25?**
Circle the correct answer.

25   250   2,500   2.5

1 mark

Reasoning Test 24 Name _____

7 Safia thinks of two whole numbers. She multiplies them together and rounds the product to the nearest 10.

Her answer is 180.

One of her numbers is 14. What is her other number?

1 mark

8 An Italian restaurant has 24 tables. Half of the tables each seat 4 people and the other half each seat 6 people.

If each person pays an average of £17 for their meal, how much money does the restaurant take in one evening when all the tables are filled?

£ _____

2 marks

Reasoning Test 24   Name _____

9 The shaded shape is a polygon but part of it is hidden.

Name four polygons that this shape could be?

Explain your answer.

2 marks

10 A boy is having his birthday party at the local play centre. His older brother and 9 friends attend the party. The play centre charges £11.75 per person for entry to the party or £15.50 per person if they supply party food.

What is the difference in price for the whole group, with and without party food?

2 marks

11 A window cleaner's large ladder has 45 rungs from top to bottom. A small ladder has 15 rungs.

The window cleaner has 12 large ladders and 8 small ladders in his garage.

How many ladder rungs are there in total on all of his ladders?

2 marks

Reasoning Test 24

Name _____

12 The table below shows how animals' ages increase compared to humans' ages. For example, a dog is the equivalent of 24 human years at age 2. Every year after age 2, a dog's age increases by 4 human years.

Animal	Animal actual years		Human equivalent years	Human equivalent years per actual year thereafter
Cat	2	=	25	+ 4
Dog	2	=	24	+ 4
Horse	3	=	6.5	+ 5
Elephant	1	=	1	+ 1

a) What age will a cat be, in human years, when it reaches age 7?

b) What age will a horse be, in human years, when it is 66.5 horse years old?

3 marks

Total marks ………/18

Reasoning Test 25   Name _____

1 What is 1,600 divided by 100?
Circle the correct answer.

| 1.6 | 160 | 0.16 | 16 |

1 mark

2 These are the times of two films showing at the local cinema.

Films	Start	Finish	Ticket price
Mary Poppins	15:35	17:48	£5.75
Secret Life of Pets 2	17:45	20:05	£6.20

How long, in minutes, is the longest film?

minutes

1 mark

3 Calculate the area of this shape.

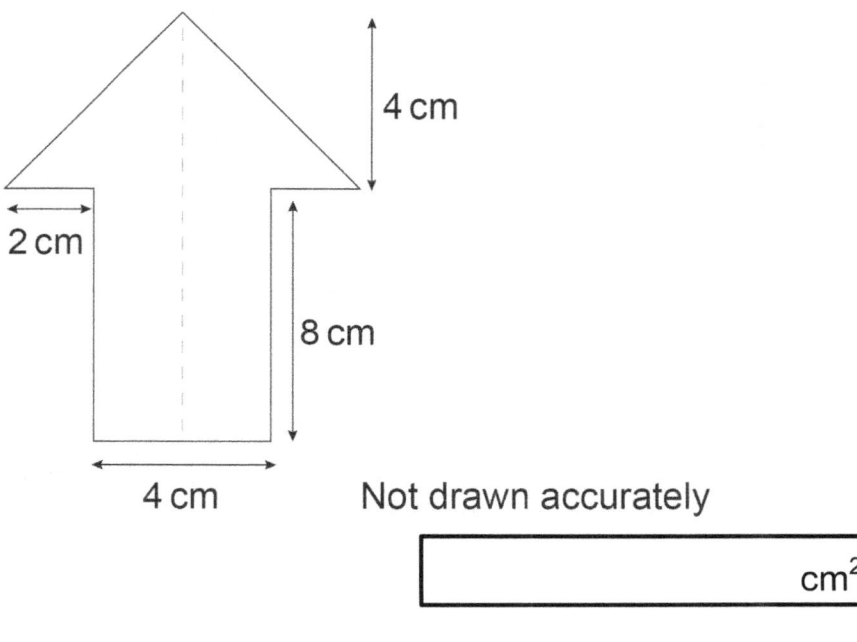

Not drawn accurately

cm²

1 mark

Reasoning Test 25 Name _____

4 What is the size of the unknown angle in this triangle?

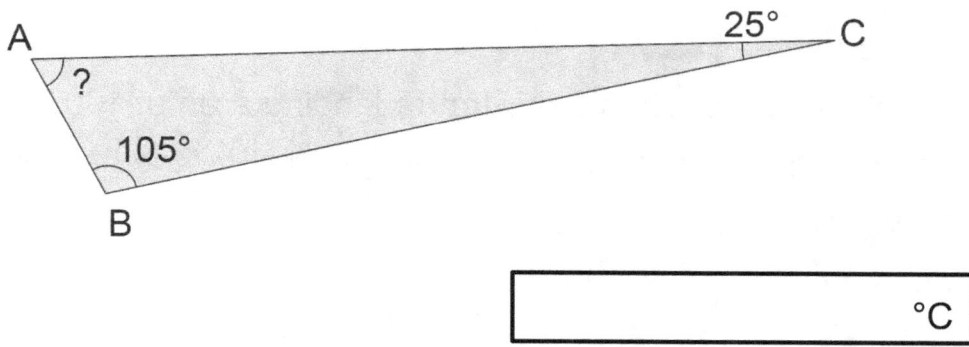

☐ °C

1 mark

5 Use the information in the bar model to find the missing fraction.

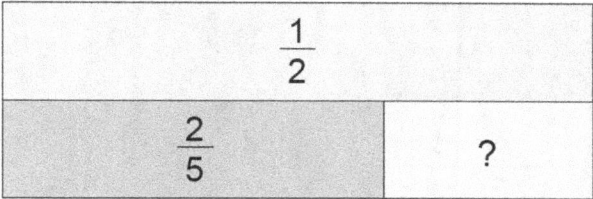

1 mark

6 What is 15 ÷ 0.2?

Circle the correct answer.

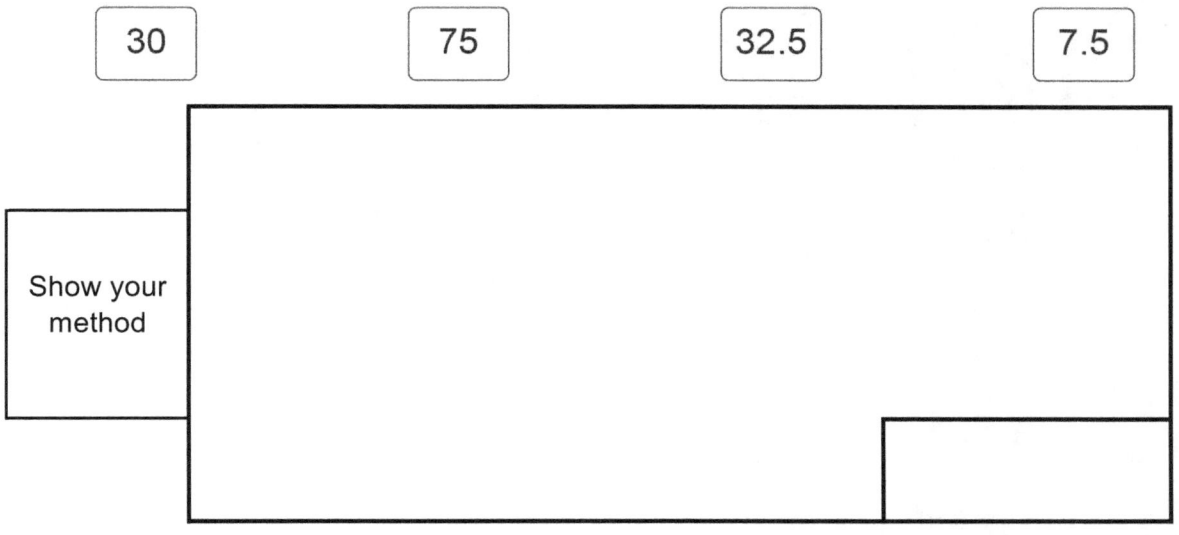

1 mark

7 A school field is 40 m × 30 m.

1 metre = 3.28 feet

An air ambulance requires a **minimum** landing area of 100 ft × 100 ft.

Is the area of the school field large enough to allow the helicopter to land? Explain your answer.

Yes / No

1 mark

Reasoning Test 25

Name _____

8 An Olympic swimmer completes 100 m breaststroke in 57.13 seconds.

If he completed another 100 m at the same speed, what would be the 200 m time?

Answer in minutes and seconds.

minutes	seconds

2 marks

9 A and B are whole numbers that total 700.

A is 150 greater than B.

What is the value of B?

You could use a bar model to help with your calculations.

700

2 marks

Reasoning Test 25 Name _____

10 A local company makes wood tables. There are two different sizes:

Small table	130 cm × 75 cm	£160.00
Large table	180 cm × 100 cm	£200.00

a) What is the area of a piece of wood for a large table top?

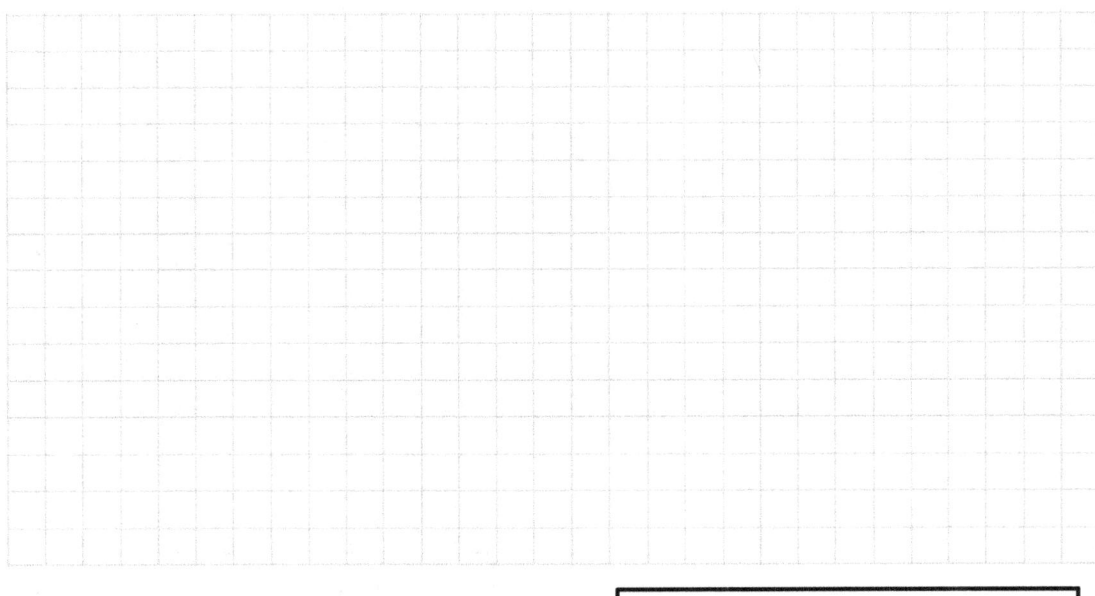

cm²

The company adds 70% profit to the cost price shown in the table.

b) If it sells 1 large and 1 small table, how much profit does it make?

£

2 marks

11 An author is writing a book. On average, she writes 7 pages per day.

If she writes every day for a whole year (365 days), how many pages will she complete?

pages

2 marks

12 On very cold days, a hotel keeps a fire burning for 12 hours. It takes 4 logs per hour to keep it alight.

a) If there are 55 very cold days in winter, how many logs will be used?

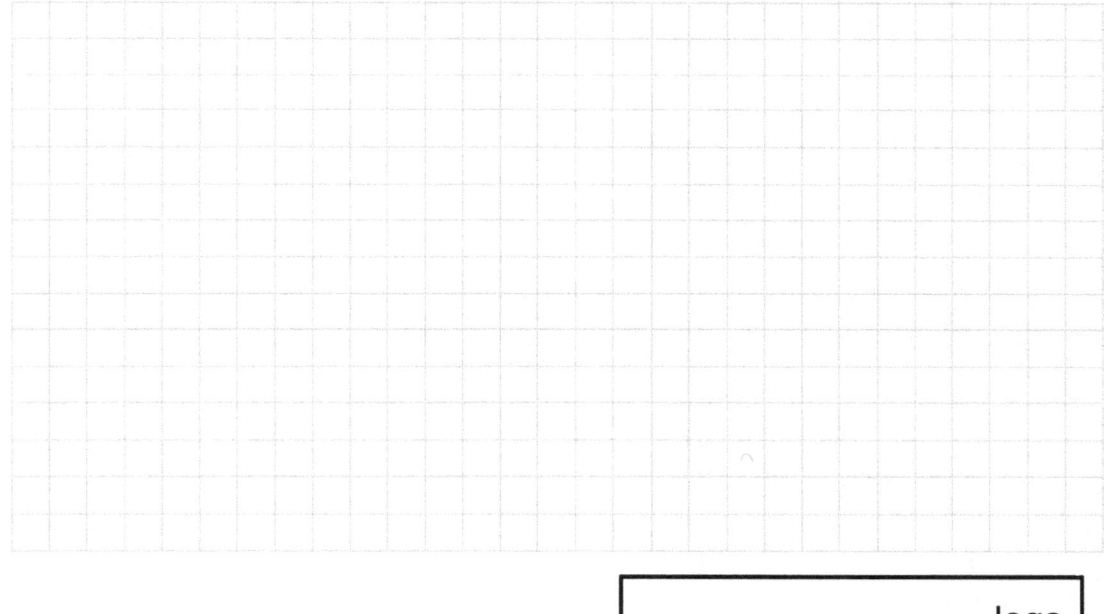

logs

Reasoning Test 25

Name _____

b) A bag of 12 logs costs £7.50. How much will 144 logs cost?

£ _____

3 marks

Total marks ………/18

Reasoning Test 26   Name _____

1 Write each number in the correct part of the Venn diagram.

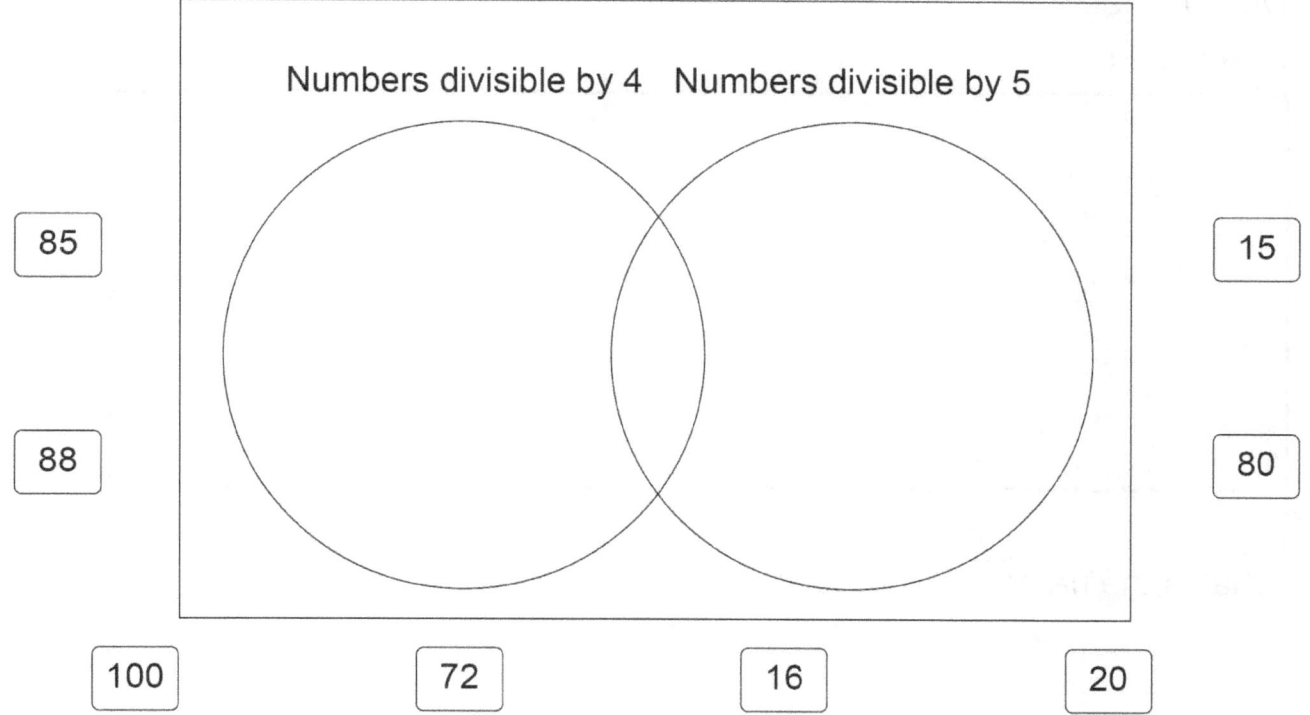

1 mark

2 This bus timetable is for the morning bus route from Wharton to Malsend.

Wharton	8.30	8.55	10.15	10.35
Ardenberry	——	9.05	10.25	10.45
Charsten	10.01	——	10.41	——
Ditton	10.23	9.43	11.03	11.23
Malsend	10.55	10.15	11.30	11.55

What is the best time to leave to get from Wharton to Malsend as quickly as possible?

1 mark

Reasoning Test 26

Name _____

3 A cuboid has length 5 cm, width 3 cm and height 2 cm.

Draw its net.

Use a ruler.

1 mark

4 What is the name for this polygon?

1 mark

5 Complete this calculation.

Write your answer as a mixed number.

$6\frac{1}{3} \times 5 =$

Show your method

1 mark

6 If 11 is 20% of 55, what percentage of 55 is 33?

1 mark

7 Elijah says that this equation is not correct.

112 × 7 = 876 ÷ 8

Explain why Elijah is right.

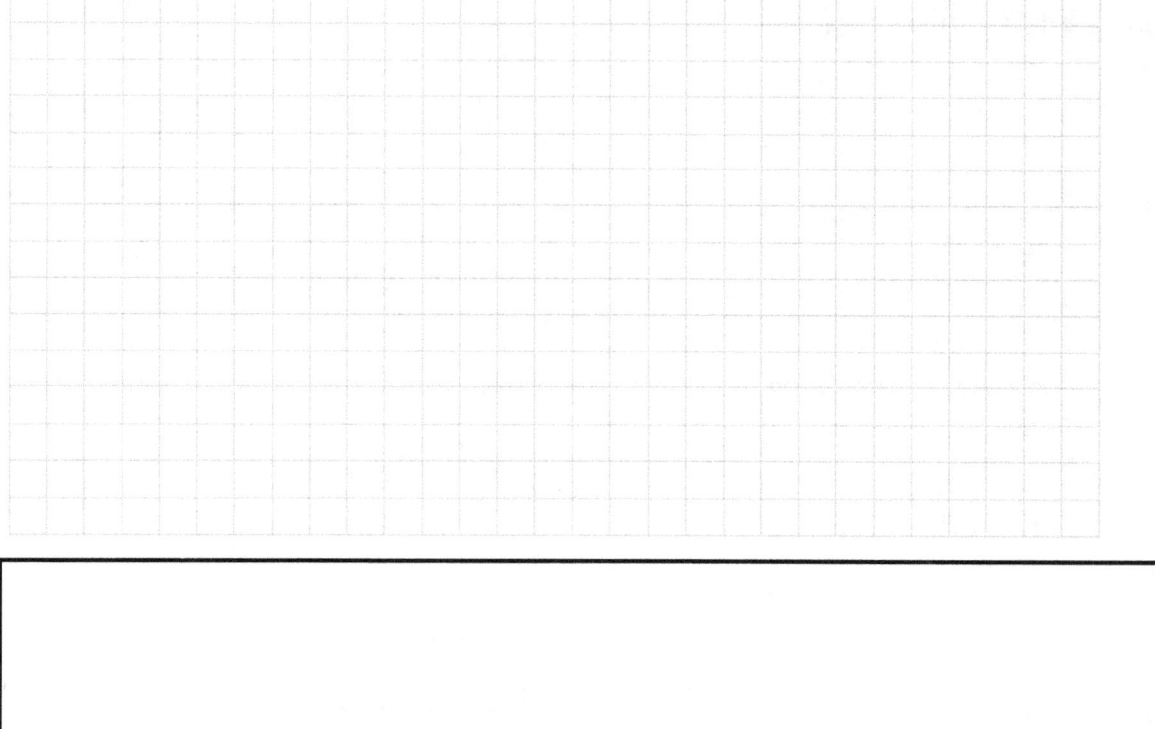

1 mark

Reasoning Test 26 Name _____

8   Ben Nevis is the highest mountain in Great Britain.
    It is 1,350 metres high.

    1 metre = 3.28 feet

    a) What is the approximate height of Ben Nevis, in feet?
       Show your working.

    _____ feet

    b) On average, a person climbs 300 m per hour. How long will it take to climb Ben Nevis at this rate?

    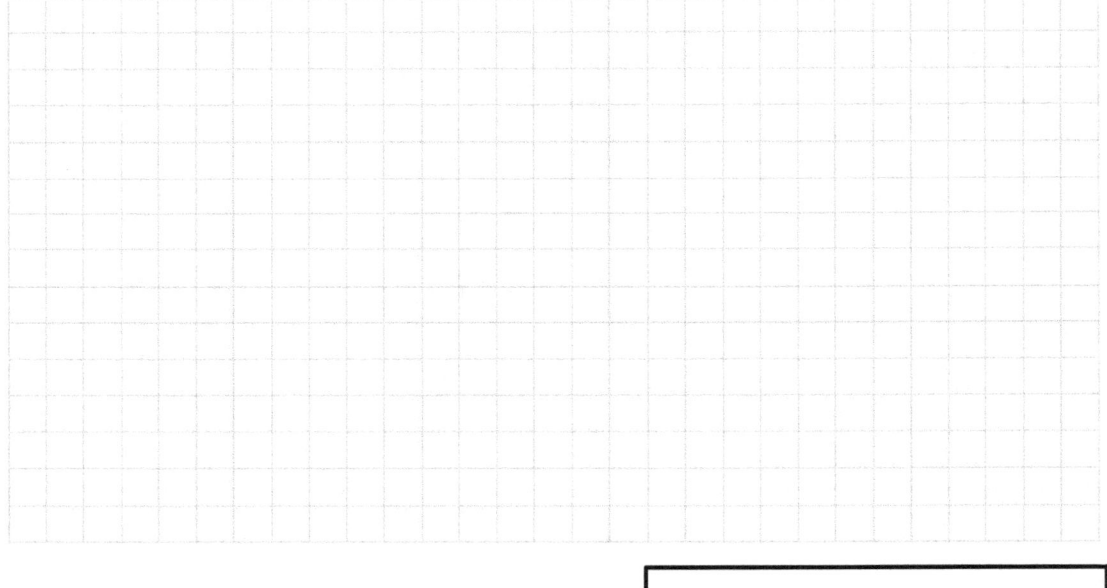

    _____ hours

    2 marks

Reasoning Test 26 Name _____

9 A school group of Year 5 children attend a swimming gala. They all compete in five rounds. Each round takes 8.5 minutes with a break of 2.5 minutes between rounds. How long does the gala last?

minutes

2 marks

10 Amir orders 99 balloons for his daughter's birthday party. She has chosen red, white and blue as the theme.

He orders the balloons and when the box arrives it is labelled: Red, white and blue balloons (2 : 4 : 5).

a) How many of each colour balloon are in the box?

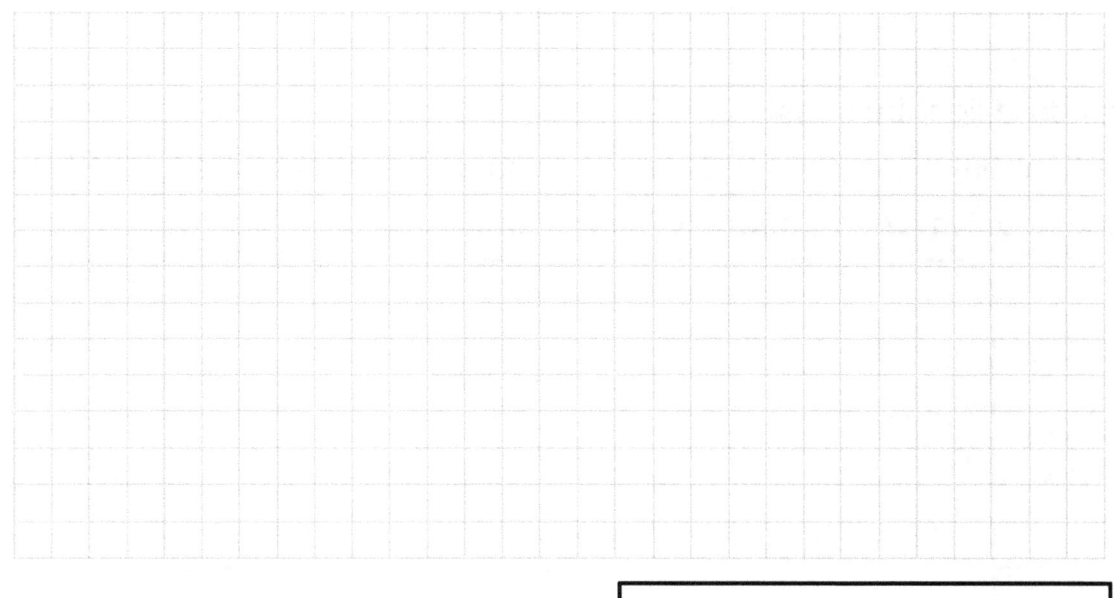

red,   white,   blue

Reasoning Test 26

Name _____

Amir's daughter's name is written on the red balloons. They are more expensive than the white and blue balloons, which are plain.

b) If the red balloons cost 12p each and the white and blue cost 9p each, how much does a box of 20 red and 50 white balloons cost?

£ ⎕

2 marks

11 A standard light bulb lasts 2,000 hours.

If the light bulb is in use for 6 hours each day, how many days will it last?

Round your answer to nearest whole number.

Show your method

_____ days

2 marks

Reasoning Test 26

Name _____

12 Jane and Paula are drawing the human figure in an art lesson.

The teacher says that the total height of the human body is approximately 7 times the height of a person's head.

Jane measures Paula's head with her ruler. Paula's head is 21.5 cm from forehead to chin.

a) How tall should she be, according to these proportions?

[ ] cm

They measure Jane's height but only have a standing measure marked in feet and inches.

2.54 cm = 1 inch
12 inches = 1 foot

b) They calculate that she is 4 ft 2 inches tall. How tall is Jane, in centimetres?

[ ] cm

3 marks

Total marks ..../18

Reasoning Test 27   Name _____

1 Circle the time that represents $\frac{1}{4}$ of 5 minutes.

   1.5 minutes   100 seconds   2 minutes   1.75 minutes   75 seconds

   1 mark

2 This table shows the amounts of prize money competitors win in a golf competition.

   a) How much prize money is given in total to the top 3 competitors?

   £ _____

   b) What is the difference between the amounts of prize money awarded for 3rd and 9th places?

   £ _____

Place	Prize (£)
1	100,000
2	75,000
3	50,000
4	10,000
5	5,000
6	1,000
7	1,000
8	1,000
9	500
10	500

1 mark

3 Draw an obtuse angle that is between 100° and 130° and an acute angle that is between 50° and 80°.

   Use a ruler and a protractor.

1 mark

Reasoning Test 27   Name _____

4   Draw the 2D front view of this shape.

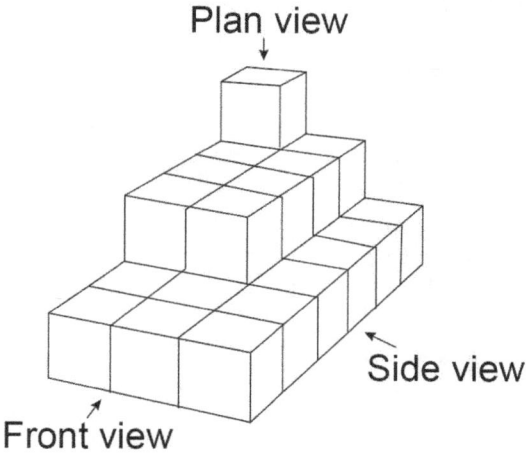

1 mark

5   Look at equations A and B. One set of scales should not balance.

Circle the set of scales that should not balance.

A

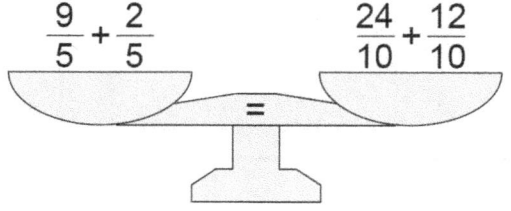

$\frac{9}{5} + \frac{2}{5}$ = $\frac{24}{10} + \frac{12}{10}$

B

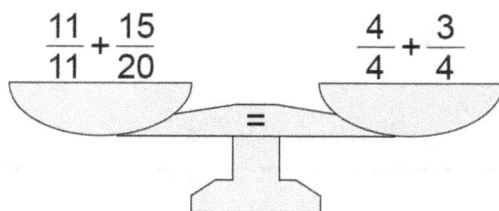

$\frac{11}{11} + \frac{15}{20}$ = $\frac{4}{4} + \frac{3}{4}$

1 mark

Reasoning Test 27   Name _____

6 Shade $\frac{1}{9}$ of this shape.

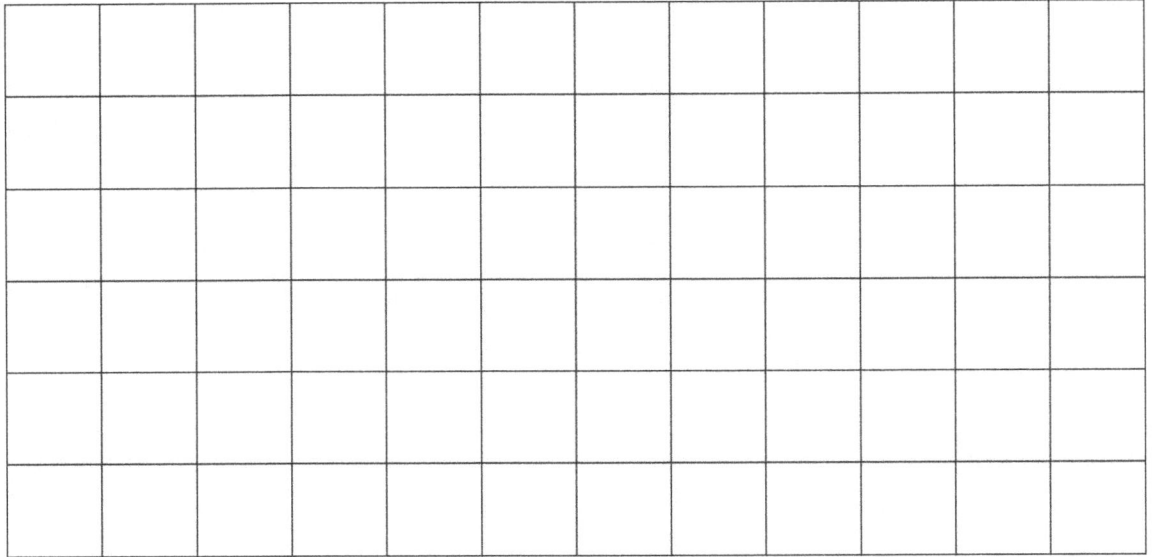

1 mark

7 How do you know that this shape is a dodecagon?

1 mark

Reasoning Test 27   Name _____

8 Alex and Brad are trying to work out who earns more money each year.
   Alex says he earns £2,200 each month. Brad earns £500 each week.
   Who earns more in a year, Alex or Brad?
   Explain how you know.

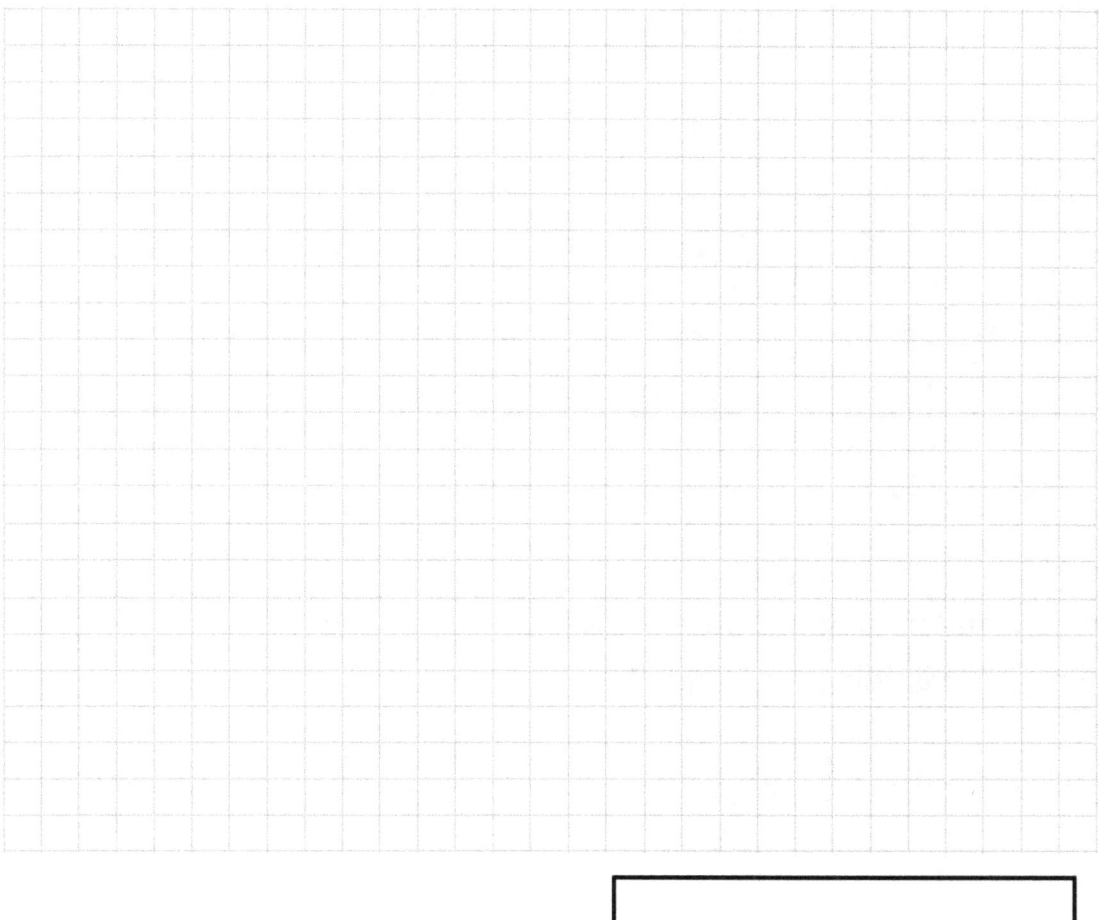

2 mark

Reasoning Test 27

9 Katie tosses three coins together. Each coin had 'heads' on one side and 'tails' on the other.

What are all the possible outcomes from Katie's three tosses?

One has been done for you.

H H H

2 marks

10 A rugby match lasts for 80 minutes.

a) How long is this in seconds?

seconds

Reasoning Test 27

Name _____

b) Each season, a rugby team plays 30 matches. How many hours of rugby does the team play throughout the season?

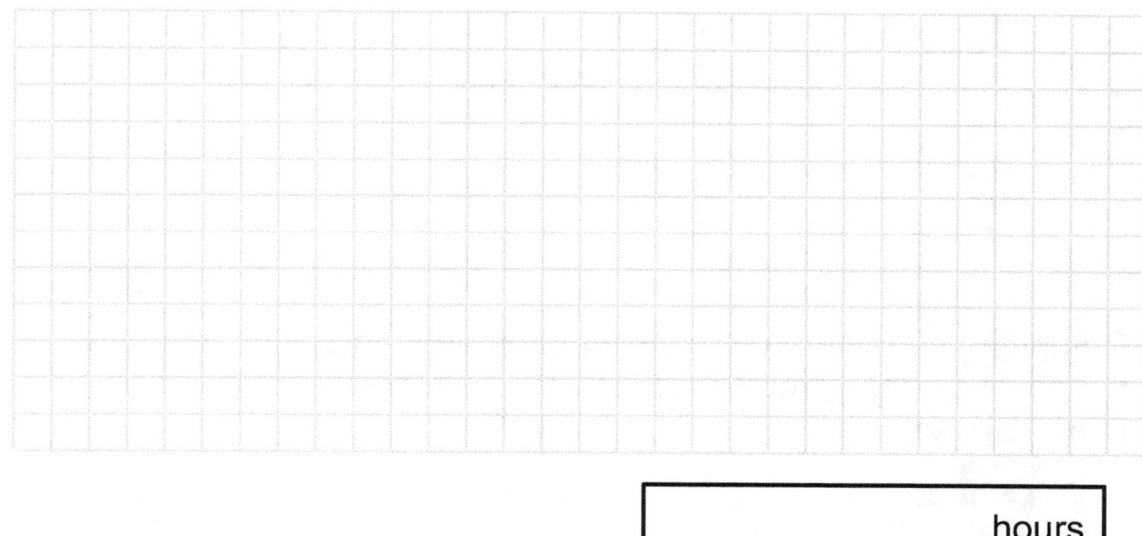

[ ] hours

2 marks

11 There are 130 cats in a shelter waiting to be rehomed. 40% are tabby, 30% are black, 20% are tortoiseshell and the rest are other colours.

a) How many cats are tabby?

Show your method [ ] cats

b) What proportion of cats are other colours?

Show your method

2 marks

Reasoning Test 27   Name _____

12 This is a pattern of squares made from match sticks.

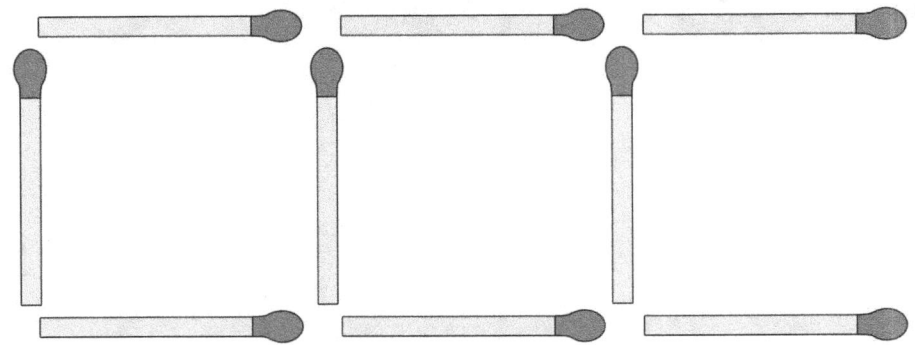

a) Complete the pattern up to six squares. Add values to the table to show how many matchsticks would be required for a certain number of squares.

Square	1	2	3	4	5	6	7	10	100
Matches	4	7	10						

b) What rule do you apply to work out how many matchsticks are required?

Show your method

3 marks

Total marks ………/18

Reasoning Test 28   Name _____

1 How many days are there in 35 weeks?
   Circle the correct answer.

   210   70   245   280   84

   1 mark

2 A group of Year 3 children measured their heights, in centimetres.
   Their results are shown in the table.

Child	Height (cm)
Seb	122
Kaitlin	118
Piran	125
Charlie	140
Jess	120

   What is the mean average height of this group of children?

   Show your method

   cm

   1 mark

3 Using a protractor and a ruler, draw a polygon with one angle of 120°.
   Your polygon must have between 3 and 8 sides.

   1 mark

Reasoning Test 28   Name _____

4 Calculate the area of this shape.

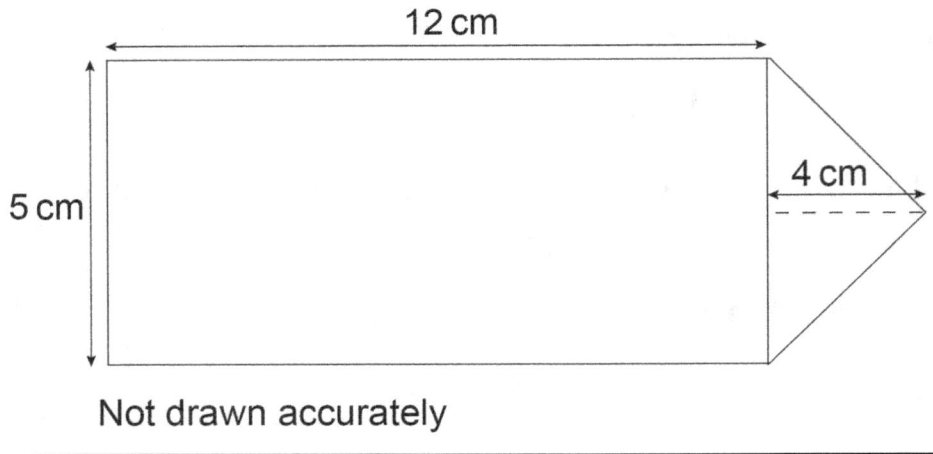

Not drawn accurately

Show your method

cm²

1 mark

5 Choose from the numbers and signs to create an inequality that is true.
Write your inequality in the boxes provided.

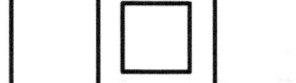

1 mark

Reasoning Test 28 Name _____

6 What is the equivalent fraction?

$$\frac{6}{9} = \frac{}{54}$$

Show your method

1 mark

7 Leah and Poppy each pick a number between 1 and 20.

Leah says, 'There's a difference of 5 between the numbers.'

Poppy says, 'If you divide my number by yours, the answer is 1.5.'

What are their numbers?

Show your method

| Leah | Poppy |

1 mark

Reasoning Test 28　　　Name _____

8　A gardener is putting 13 m of a wooden border around some flowerbeds.
　　The wood comes in 0.75 m and 1.25 m lengths.

　　If the gardener uses four 0.75 m lengths, how many 1.25 m lengths will she need to complete the border?

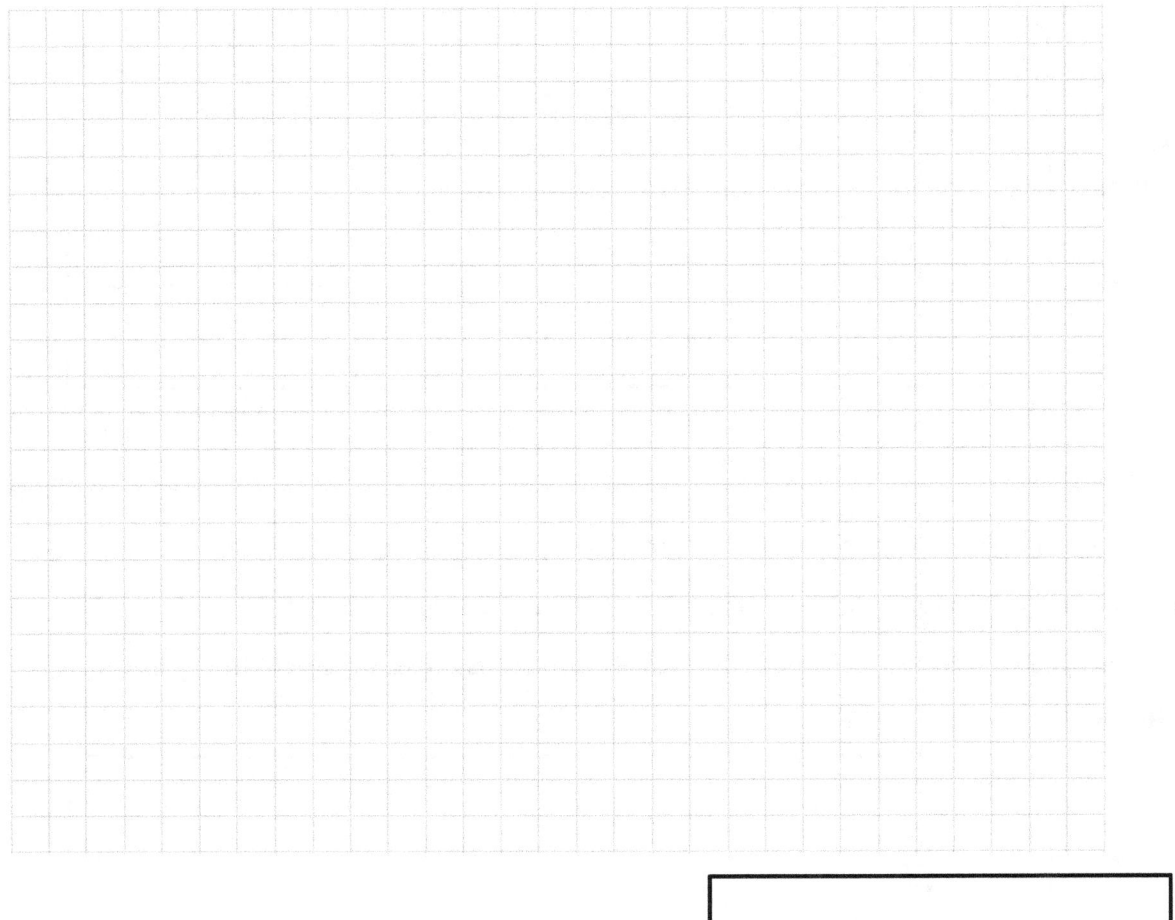

2 marks

Reasoning Test 28   Name _____

9 A shape is made up of a square and rectangle.

The perimeter of the shape is 64 m and the area of the square is 64 m².

What is the area of the black rectangle?

Show your method

m²

2 marks

10 During July, the temperature at 6 p.m. in Spain is 28 °C on average. At night, the temperature drops 0.5 degrees every hour from 6 p.m. until 12 midnight and then by 2 degrees every hour until 4 a.m.

a) What is the temperature at 4 a.m.?

°C

Reasoning Test 28   Name _____

The temperature is 17 °C at 6 a.m. and for the next 2 hours, it rises at 3 degrees per hour.

b) How many degrees must it rise in the remaining hours to reach 35 °C by midday?

[_____] degrees

2 marks

11 There are 725 children in a school. The ratio of boys to girls is 2 : 3.

a) How many boys are there in the school?

Show your method

[_____] boys

Reasoning Test 28   Name _____

In KS4, there are 400 children. $\frac{3}{5}$ of the children are girls.

b) How many girls are there in KS4?

2 marks

12 At a restaurant, a couple choose dinner from the specials menu. They have a budget of £30.

a) What choices of food could they make to stay within budget if they both have a starter, main and dessert?

Use the table to support your calculations.

Food	Maria	James
Starter		
Main		
Dessert		
Total		

January specials

Starters

Garlic bread ~ 2.95

Soup of the day ~ 3.95

Mains

Grilled haddock ~ 7.50

Steak frites ~ 6.95

Dessert

See board ~ 3.50

Reasoning Test 28   Name _____

A family of four have dinner at the restaurant. Their bill comes to £65.80.
They have a voucher for 15% discount on the price of their meal.

b) How much do they pay for their meal?

Show your method

3 marks

Total marks ..../18

Reasoning Test 29   Name _____

1 Match each mass in kilograms to the equivalent mass in pounds.
Draw arrows between the boxes.

65 kg		2.2 lb
10 kg		33 lb
1 kg		22 lb
15 kg		143 lb

1 mark

2 Four KS2 classes participated in a sponsored walk to raise money for charity.
The pie chart shows the results of their fundraising.

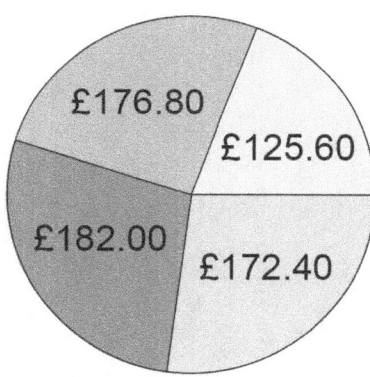

Sponsored walk

£176.80 £125.60 £182.00 £172.40

☐ Class 3   ☐ Class 4   ■ Class 5   ☐ Class 6

How much more money did Class 4 raise than Class 3?

Show your method

£

1 mark

3 This shape has an area of 36 cm² and a perimeter of 30 cm.

Draw a rectangle with the same area but a different perimeter.
Use a ruler.

1 square = 1 cm × 1 cm

1 mark

4 Plot these coordinates on the graph and then join the points, in order, to make a polygon.

Name the polygon.

(3, 2) (6, 2) (6, 6) (3, 4)

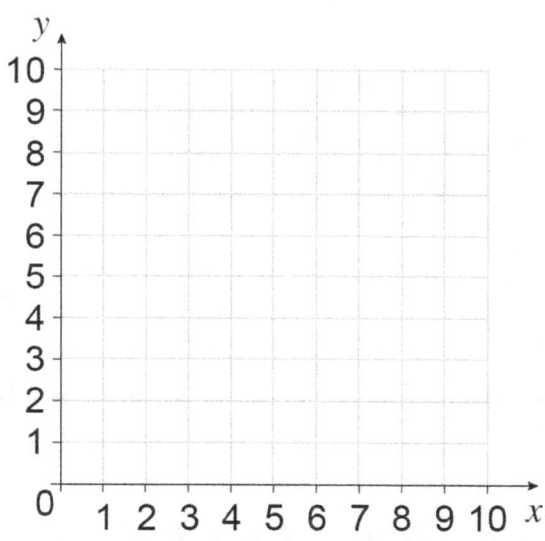

Polygon = ▢

1 mark

5 What is 70% of 60 added to 40% of 90?

Show your method

1 mark

Reasoning Test 29

Name _____

6 Add the fractions.

Give your answer as a mixed number.

$\frac{4}{5} + \frac{2}{3} =$

Show your method

1 mark

7 Aria has wanted a new jumper for a while. She has found one in the sales in two different shops.

The non-sale price in shop A is £35 and in shop B, the non-sale price is £29.

She has £20 to spend. Is the price of either jumper less than £20 in the sale?

Explain your answer.

40% off all knitwear	30% off all clothes
Shop A	Shop B

Show your method

Yes / No

1 mark

8 A bottle of children's medicine holds 200 ml.

Child's age	Dose	How often (24 hours)
3–6 months	2.5 ml	Up to 4 times
6–24 months	5 ml	Up to 4 times
2–4 years	7.5 ml	Up to 4 times
4–6 years	10 ml	Up to 4 times

a) How many doses will the bottle provide for a child aged 4–6 years?

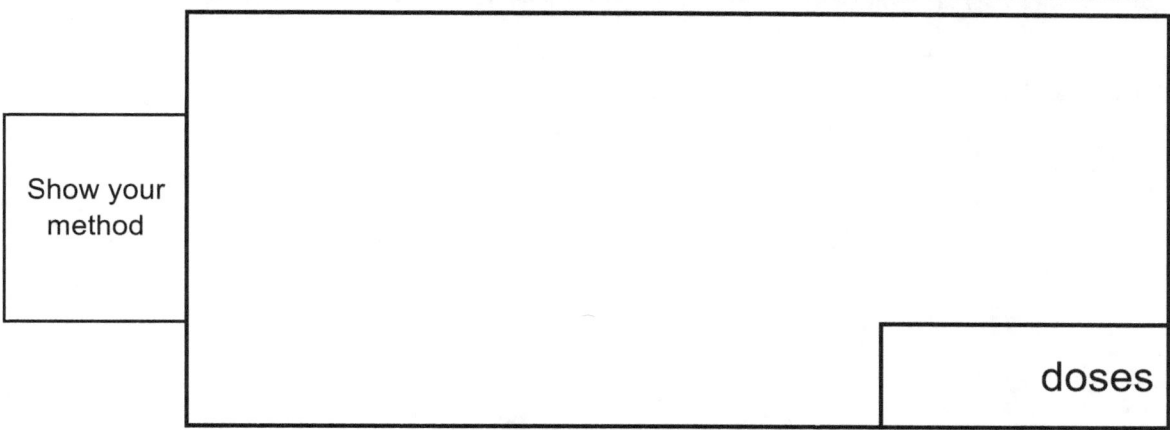

Show your method

doses

b) How many doses will the bottle provide for a child aged 2–4 years? Round your answer to the nearest full dose.

Show your method

doses

2 marks

Reasoning Test 29

Name _____

9 Lily needs to build a tower of cubes 39 cm high.

She has 2.5 cm cubes and 1.8 cm cubes.

If she uses five of the 1.8 cm cubes, how many 2.5 cm cubes will she need?

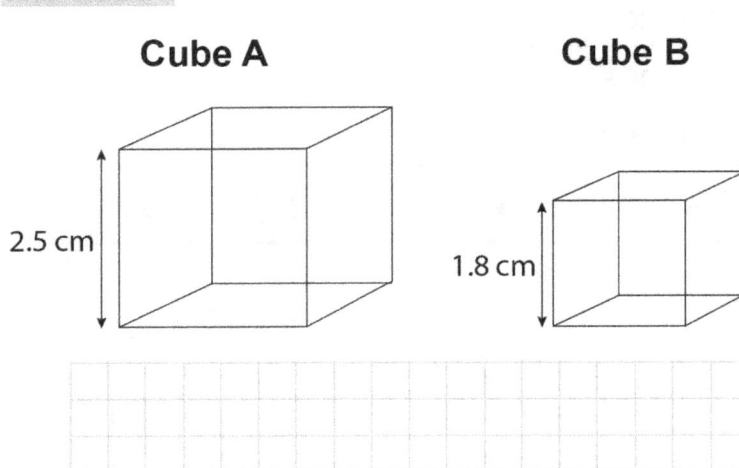

of cube A

2 marks

Reasoning Test 29

10 These numbers are written in Roman numerals.

Find their product. Write your answer in Roman numerals.

XXVIII × XLVII = ☐

2 marks

Reasoning Test 29   Name _____

11 Henry draws a regular quadrilateral with an area of 81 cm².

a) What is the perimeter of this quadrilateral?

Show your method

[   ] cm

Henry enlarges the lengths in the shape by a scale factor of 3.

b) What is the area of the new shape?

[   ] cm²

2 marks

Reasoning Test 29 Name _____

12 Suki buys a smoothie and a flapjack. Together they cost £3.50.

Jake buys three smoothies and two flapjacks, costing £9.25 altogether.

a) How much do smoothies and flapjacks cost on their own?

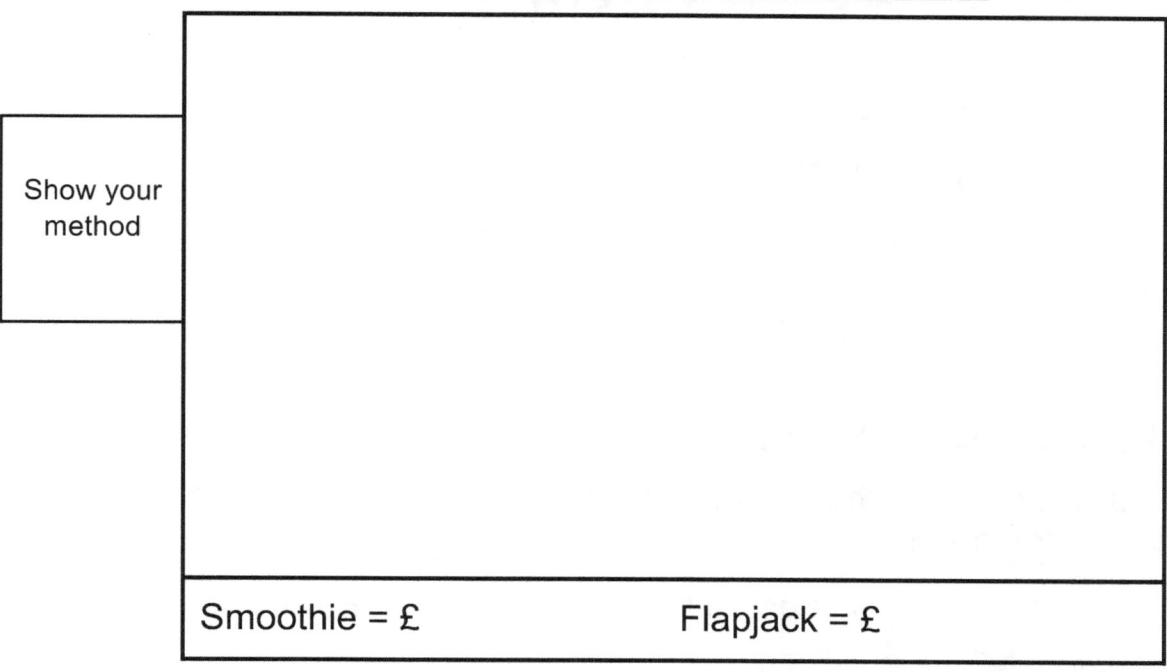

Smoothie = £   Flapjack = £

Cookies cost £1.50 in the supermarket. A wholesaler buys cookies for 20% of their price in the supermarket.

b) If the wholesaler buys 700 cookies at this lower price, how much would these cookies cost in total?

£

3 marks

Total marks ..../18

Reasoning Test 30   Name _____

1. Circle the number that is represented by these Roman numerals.

   XXXIV

   | 66 | 36 | 34 | 25 | 306 |

   1 mark

2. An author has written two best-selling children's novels.

   This bar chart shows the numbers of each book sold per month worldwide, from January to April.

   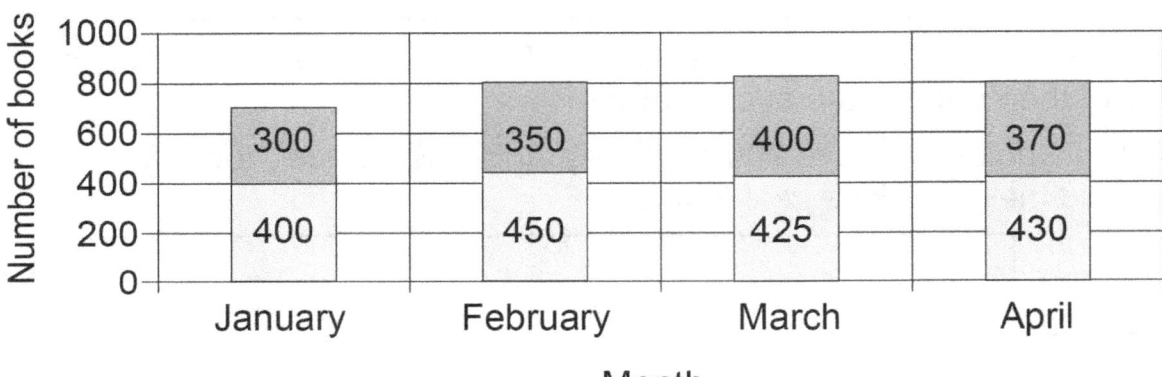

   What is the total number of books sold in March and April?

   | Show your method | | books |

   1 mark

3 Each cube weighs 5.5 g.

How much do these cubes weigh in total?

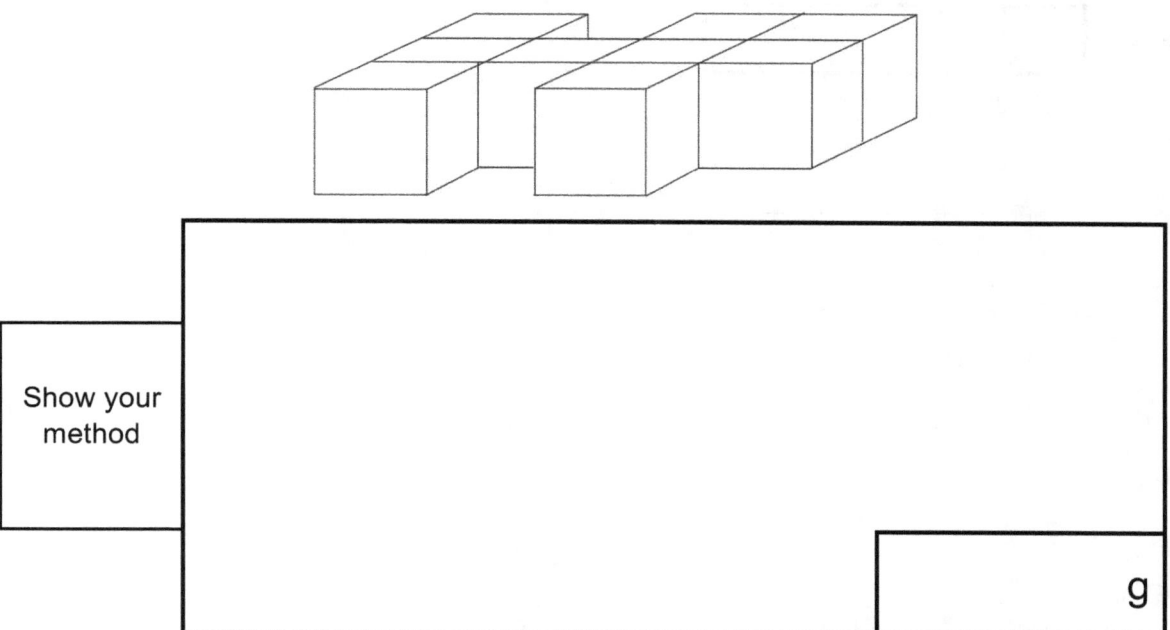

Show your method

g

1 mark

4 Rotate this shape 270° clockwise about point A. Draw the shape in its new position.

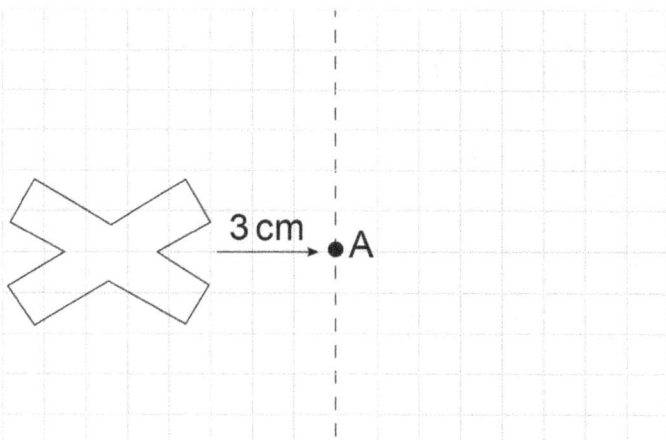

1 mark

5 Use this shape to draw the correct fractions and work out the answer to this calculation.

$\frac{1}{3} \div 4 = \boxed{}$

1 mark

6 Complete this calculation with fractions.

$\frac{5}{8} - \frac{1}{3} = \boxed{}$

1 mark

Reasoning Test 30   Name _____

7   Tom and Ravvi have a competition to see how far they can skate in an hour.
    They send each other messages at the end, sharing their results.
    Who went further?

    > 8 km = 5 mile approximately
    > 1 mile = 1,600 m approximately

    > I made it to the blue house and back. It was 10.5 km
    > – Tom

    > 6 miles exactly!
    > Yes ☺ R xx

    Show your method

    1 mark

8   Letitia is making coffee at the school fete. For every 250 ml cup of coffee, she uses 1 teaspoon (5 grams) of coffee powder. She charges 50p per cup.

    a) How much coffee powder does she use if she makes a 10-litre vat of coffee?

    Show your method

    ___ g

Coffee costs £3.50 per 100 g.

b) If she buys 300 g of coffee powder, how many cups of coffee does she need to sell, to cover the cost?

cups

2 marks

9 What is the value of x when y = 18?

$$2(2x + y) = 84$$

x =

2 marks

Reasoning Test 30

Name _____

10 The school swimming pool is 25 metres long. Idris and Kala have been selected for time trials to qualify for a regional competition. They must each swim a length in the expected qualifying time for their age category.

a) What is the average length time for each of the boys?

	Age category	Lengths	Total time (seconds)	Qualifying time (seconds per length)
Idris	< 10	8	160	< 20.5
Kala	< 12	14	210	< 18.5

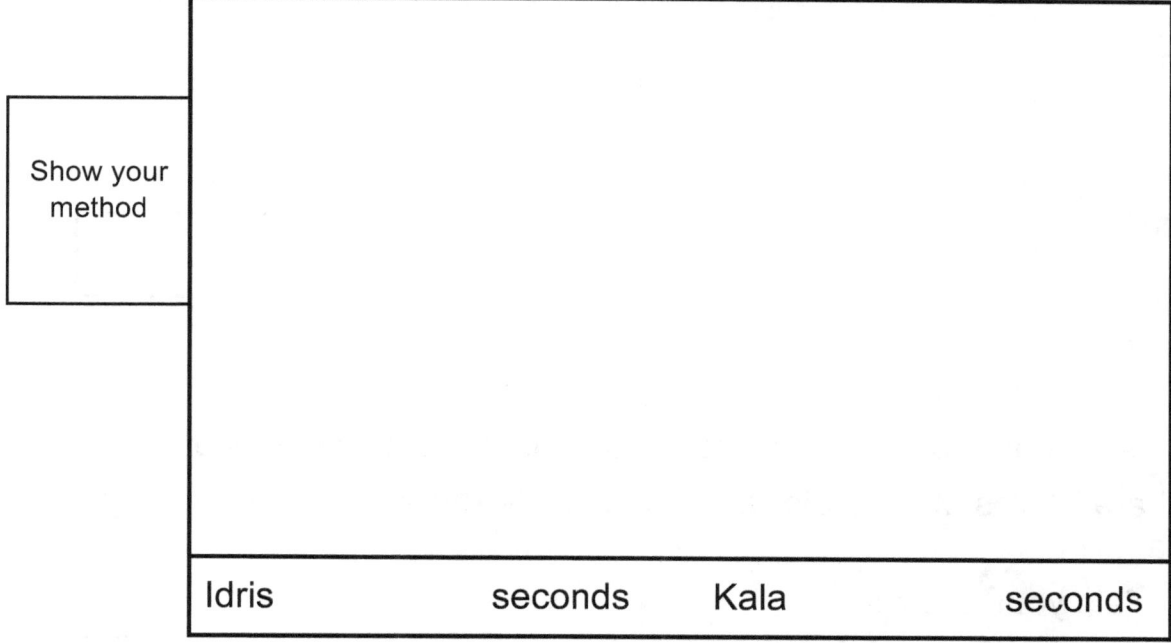

Show your method

Idris _____ seconds   Kala _____ seconds

b) Do they **both** qualify?

Yes / No

2 marks

Reasoning Test 30

Name _____

11 A volunteer group is running an event for children in local hospitals. They need 70% of their 480 volunteers to attend for the event to go ahead.

If 324 people have volunteered, are there enough people to run the event?

Explain your answer.

Yes / No

2 marks

12 It takes 3 hours 45 minutes to travel to London from Manchester.

a) If Lara leaves Manchester at 5.40 a.m. will she arrive in London by 9 a.m.?

Explain your answer.

Show your method

Yes / No

Reasoning Test 30

Name _____

For every 15 minutes of travel, Lara's employer pays her £8.46.

b) Her train ticket costs £126.00. How much money will she have left after paying her train fair?

£ _____

3 marks

Total marks ………/18

Reasoning Test 1: Mark Scheme

Question	Requirement	Mark	Additional guidance	Content domain reference
1	32, 40, 56	1	All three answers must be correct to achieve the mark.	3N1b
2a	210	1		3S1
2b	130			
3	7 hours 30 minutes	1	Do not accept 7.5 hours.	4M4b
4	Square, pentagon, hexagon and heptagon must all be ticked.	1		4G2a
5	$\frac{4}{16} < \frac{5}{8} < \frac{3}{4}$ $\frac{4}{16} < \frac{7}{14} < \frac{3}{4}$ $\frac{4}{16} < \frac{7}{14} < \frac{5}{8}$ $\frac{7}{14} < \frac{5}{8} < \frac{3}{4}$	1		5F4
6	$\frac{2}{8}, \frac{1}{2}, \frac{3}{4}, \frac{7}{8}, \frac{4}{2}$	1		5F4
7	No	1	Children should work out the following calculations: 2,054 + 1,880 = 3,934 5,000 − 3,934 = 1,066 1,066 is less than 1,200 and therefore there are not enough pieces of wood. Children must show that 1,066 < 1,200 to be awarded the mark.	5C2
8	£8.30	2	Award **1 mark** for an incorrect answer showing the correct method. 9.50 + 9.50 = 19.00 7.50 + 7.50 + 7.50 = 22.50 22.50 + 19 = 41.50 41.50 ÷ 5 = 8.30 correct answer: **2 marks** 41.50 ÷ 5 = incorrect answer: **1 mark**	6C8
9	96°	2	Award **1 mark** for an incorrect answer showing the correct method. 42 + 42 = 84 180 − 84 = incorrect answer: **1 mark** 180 − 84 = 96° = correct answer: **2 marks**	6G4a

Reasoning Test 1: Mark Scheme

10	Yes	2	Award **1 mark** for an incorrect answer showing the correct method. 120 + 132 = 252 35 + 35 + 35 = 105 350 − 105 = incorrect answer: **1 mark** 350 − 105 = 245 correct answer: **2 marks**	6C8
11a	440 g	1	Award **1 mark** for each correct answer or **1 mark** for two correct methods with one correct and one incorrect answer. $\frac{2750}{250} = 11$ 11 × 40 = 440 g correct answer: **1 mark**	6R2/6M5
11b	412.5 ml	1	$\frac{2750}{10} = 275$ $\frac{275}{2} = 137.5$ 275 + 137.5 = 412.5 correct answer: **1 mark**	
12	2,015 m²	3	**1 mark** for calculating 108 × 45 = 4,860 100 × 45 = 4,500 8 × 45 = 360 4,500 + 360 = 4,860 correct multiplication: **1 mark** 6,875 − 4,860 = incorrect answer: **1 mark** 6,875 − 4,860 = 2,015 correct answer: **3 marks**	6M7a/6M9
Total		18		

Reasoning Test 2: Mark Scheme

Question	Requirement	Mark	Additional guidance	Content domain reference
1	7, 10, 13	1		4N5
2	34°C	1		3S2
3	32 cm	1		4M7a
4	Coordinates join to make a parallelogram.	1	Must be a parallelogram not a rhombus.	4P3b
5	$\frac{21}{28}$, $\frac{75}{100}$, $\frac{18}{24}$	1		4F2
6	Apples = £2 Pie = £3 Box of chocolates = £4 Bottle of water = £1	1		4F10b
7	Any three answers from: • It has more than four sides. • It is not a quadrilateral. • It does not have a set of parallel sides. • It has more than four angles.	1		5G2b
8a	£2	1	Award **1 mark** for correct working and one correct answer.	6R1
8b	8	1	Award **1 mark** for correct working and one correct answer.	
9	$3b - 7 = 29$ The value of *b* in equation **B** is greater. Equation **A** $22 - 6 = 16$ $4 \times b = 16$ $16 \div 4 = 4$ $b = 4$ Equation **B** $29 + 7 = 36$ $3 \times b = 36$ $36 \div 3 = 12$ $b = 12$ When calculating both values of *b* it is seen that in equation **A**, $b = 4$ and in equation **B**, $b = 12$. 12 is greater than 4. So *b* is worth more in equation **B**.	2	Award **1 mark** where calculations are both correct yet answer is incorrect.	6A1

Reasoning Test 2: Mark Scheme

10a	5 hours	1	Award **1 mark** for correct working and one correct answer.	6M9
10b	7 hours	1	Award **1 mark** for correct working and one correct answer.	
11	6,125 ml yellow and 875 ml blue	2	Award **1 mark** for correct working and incorrect answer.	6R4
12a	6	2	Award **2 marks** for correct working and one incorrect answer.	6F11/6R1
12b	$\frac{1}{16}$	1		
Total		18		

Reasoning Test 3: Mark Scheme

Question	Requirement	Mark	Additional guidance	Content domain reference
1	4, 25, 49	1		3N2a/5C5d
2	£168	1		4S2
3	67.5 cm²	1		6M7b
4	Rectangle	1		4M7a
5	800 ml	1		5M5
6	0.375	1		6F6
7	True. The sum of three even numbers can never make an odd number because adding two even numbers gives an even number, then adding another even number gives an even number as well, so the answer will always be an even number that is divisible by 2.	1	Children should demonstrate an understanding of even and odd numbers being relevant to the answer. An example like this may be seen: For example, 2 + 4 + 6 = 12 Even numbers can all be divided by two. If three even numbers are added they are still divisible by 2, meaning they are not odd. Odd numbers cannot be halved as there is always an odd number at the end for example, 7 = 2 + 2 + 2 + 1.	5C4
8a	£1.80	1	Award **1 mark** for correct working in both parts but only one correct answer.	5M9a
8b	£4.50	1	Award **1 mark** for correct working in both parts but only one correct answer.	
9	32 m	2	Award **1 mark** for demonstrating halving and finding a third, working backwards and making a table. Award **1 mark** for incorrect answer of 16 cm with correct method. Award **2 marks** for correct answer of 32 m.	6R1/6M7a
10a	3.2 ml	1	Award **1 mark** for correct methods and only 1 correct answer.	6S3/6F9a
10b	0.9125 L	1	Award **1 mark** for correct methods and only 1 correct answer.	
11a	10.5	1	Award **1 mark** for correct methods and only 1 correct answer.	6R1
11b	12,450 m	1	Award **1 mark** for correct methods and only 1 correct answer.	
12a	492,700	1	Award **2 marks** for correct method and one incorrect answer.	6F11/6C8
12b	£295,620 profit	2	Award **2 marks** for correct method and one incorrect answer.	
Total		18		

Reasoning Test 4: Mark Scheme

Question	Requirement	Mark	Additional guidance	Content domain reference
1	56, 0.8, 70	1		4c6b
2	371 minutes	1		3S1/4M4c
3	38 mm	1		3M9b
4	b and c ticked	1		5G3b
5	$1\frac{9}{10} = \frac{19}{10}$ $2 = \frac{18}{9}$ $2\frac{4}{8} = \frac{20}{8}$ $1\frac{1}{2} = \frac{6}{4}$	1		5F2b
6a	2 out of 3 triangles	1		5F2b
6b	8 out of 12 squares shaded			
6c	10 out of 15 squares			
7	Jack is correct with 1,170.	1	Children must show the calculations and working such as: $78 \times 15 = 1,170$ $78 \times 10 = 780$ $78 \times 5 = 390$ $780 + 390 = 1,170$	5C7a
8a	1,440 km	1	Award **1 mark** for correct methods and 1 correct answer. Award **1 mark** for two correct methods.	5C7a/5C8a
8b	520 km	1	Award **1 mark** for correct methods and 1 correct answer. Award **1 mark** for two correct methods.	
9a	17	1		6A2
9b	$6n - 8$	1		
10	£33	2	Award **1 mark** for two correct methods and no correct answers. Award **1 mark** for 1 correct answer.	6C8/6R2
11a	40	1	Award **1 mark** for correct working and no correct answers, for example $\frac{320}{4}$ = incorrect answer	6R4
11b	5p	1		
12a	1,230 minutes	3	Award **1 mark** for correct working and no correct answers. Award **1 mark** for one correct answer.	6C7a/6C8
12b	£31,500			
Total		18		

Reasoning Test 5: Mark Scheme

Question	Requirement	Mark	Additional guidance	Content domain reference
1	4, 12, 6	1		3C4
2	20, 21, 22, 23, 24, 25 and 26 July 2019 10, 11, 12, 13, 14, 15, 16 and 17 August 2019	1		3S1
3	Shapes b, c and e should be ticked. Square, equilateral triangle, hexagon	1		5G2B
4	(0, 0) (0, 6) (12, 0) (12, 9)	1		5P2
5	$\frac{5}{20} + \frac{12}{20} = \frac{17}{20}$	1		5F2a
6	£315	1		5F10
7	Arjun runs faster because he is taking 5 minutes to run each kilometre.	1	Children should calculate the number of minutes to run each kilometre to compare. $\frac{54}{10}$ = 5.4 minutes $\frac{40}{8}$ = 5 minutes Therefore 5 minutes per kilometre is faster and Arjun runs at this faster pace.	4c6a/5C6b
8a	£27.20 more per week	1	5.90 × 16 = 94.40 4.20 × 16 = 67.20 Difference = £27.20	5F10/5C8a
8b	£123.20	1	7.70 × 16 = 123.20	
9	Hexagon = 35	2	35 + 35 + 10 + 10 = 90 35 + 40 = 75 35 + 40 + 10 = 85 Award **1 mark** for evidence of two of these correct calculations.	6A5
10	3 kg	2	2 hours 55 − 25 extra minutes = 2 hours 30 minutes 2 hours 30 minutes = 150 minutes 25 × 6 = 150 minutes 6 × 500 g = 3,000 g or 3 kg Award **1 mark** for evidence of working to reach 150 minutes with the incorrect answer.	6R1/5M4
11a	90 g	1	She has $\frac{3}{4}$ of the amount of cream, applying this to the chocolate: $\frac{3}{4}$ of 120 = 90.	6R4/5F12
11b	875 ml	1	500 ÷ 4 = 125 ml 125 × 3 = 375 500 + 375 = 875 ml	

Reasoning Test 5: Mark Scheme

12a	60 red cars	3	150 seconds = 2.5 minutes $\frac{30}{2.5} = 12$ 12 × 5 = 60 red cars	6R4/6C4
12b	140 white and 40 black cars		90 seconds = 1.5 minutes 20 × 1.5 = 30 7 × 20 = 140 white cars and 2 × 20 = 40 black cars	
Total		18		

Reasoning Test 6: Mark Scheme

Question	Requirement	Mark	Additional guidance	Content domain reference
1	21, 28	1		3C8
2a	35%	1		3S2
2b	30 people			
3	Triangular prism: 5, 9, 6 Square-based pyramid: 5, 8, 5 Octahedron: 8, 12, 6	1		3G3b
4	105°	1		4C4/5G2a
5	$\frac{4}{5}$ of 120 is greater as 96 is greater than 90.	1		4F4
6	$\frac{2}{8}, \frac{9}{18}, \frac{3}{4}, \frac{8}{8}, \frac{4}{2}$	1		5F3
7	Mount-Harford Because 12 × 2.5 is 30 5.8 × 4 = 23.2 Therefore 30 denotes more rainfall.	1		5M5
8a	4.40 p.m.	1		5C4
8b	8.10 p.m.	1		
9	Because you are adding 10 and dividing by 2, which is the same as adding 5.	2	Award **1 mark** where pupils demonstrate an understanding of one operation cancelling the other out.	5C8a
10a	1,280 bears	1		5C7a
10b	£720	1		
11a	There are 405 sheep and 27 ducks.	1		6R1/6M7a
11b	4 km × 5 km, 2 km × 10 km, 1 km × 20 km	1		
12	22 sweets	3	Award **1 mark** for attempt to work backwards starting with 1 sweet. Award **2 marks** if all steps are correct with incorrect final answer.	6C8
Total		18		

Reasoning Test 7: Mark Scheme

Question	Requirement	Mark	Additional guidance	Content domain reference
1	16 + 56 = 72 or 32 + 40 = 72 or 24 + 48 = 72	1		3NB1
2a	13	1		3S2
2b	76			
3	Line of exactly 58 mm (5.8 cm)	1		3M9b
4	18 faces	1		5G3b
5	$\frac{5}{40} = \frac{1}{8}$	1		5F5
6	Any two of $\frac{40}{100}$, $\frac{4}{10}$, 0.4, $\frac{16}{40}$	1		4F6b/6a
7	Declan is correct: 17 **is** a prime number as it is greater than 1 and has no factors apart from 1 and itself.	1		5C5a
8	12 × 14 = £168	2	**Award 1** mark for the correct calculation e.g. 6 × 10 = 60 6 × 4 = 24 14 × 12 = incorrect answer	5M9a
9	Any two of: 60, 64, 68, 72, 76, 80, 84, 88, 92, 96.	2		6C4
10a	180 cm long × 180 cm wide	1		6C8/6M7c
10b	180 cm long × 240 cm wide	1		
11a	60 girls	1		6R1/6R4
11b	60 boys	1		
12a	10.20 a.m.	3	**Award 1** mark for one correct answer. **Award 2 marks** for one correct answer and one correct calculation with incorrect answer.	6C7a/6M9
12b	£44.20			
Total		18		

Reasoning Test 8: Mark Scheme

Question	Requirement	Mark	Additional guidance	Content domain reference
1	12, 22, 1, 125, 67	1		4N4b
2a	0.8 g	1		4S2
2b	500 ml			
3	Parallelogram *a* and *c* Triangle none circled Pentagon *b* and *d*	1		4G4
4		1	Shape must move into lower left quadrant.	5P2
5	$\frac{20}{12} = 1\frac{8}{12}$ or $1\frac{2}{3}$	1		5F4
6	$2.2 = 2\frac{2}{10}$ or $2\frac{1}{5}$	1		5F2a
7	There is no integer that you can multiply by itself that will give 32. 16 has an odd number of factors (1, 2, 4, 8, 16) which shows it is a square number, whereas 32 has an even number of factors (1, 2, 4, 8, 16, 32).	1	Children should explain that 16 has an odd number of factors (1, 2, 4, 8, 16) which shows it is a square number, whereas 32 has an even number of factors (1, 2, 4, 8, 16, 32). This is a key property of square numbers.	5C5d
8a	8	1		5C7b/6C7a
8b	£161.25	1		
9a	5	1		5C5a/6R4
9b	60	1		
10a	78	1		5C6a/5C7a
10b	7.8 × 5 = 39 m	1		
11a	£6	1		6R1/6C4
11b	£25	1		
12a	208 m	3	**Award 1** mark for one correct answer. **Award 2 marks** for one correct answer and one correct calculation with incorrect answer.	5M7A/6C8
12b	£1,225			
Total		18		

Reasoning Test 9: Mark Scheme

Question	Requirement	Mark	Additional guidance	Content domain reference
1	0.5	1		4C6b
2a	Clock face at 12.15	1		3M4a/4M4a
2b	12:15			
3	Children draw a 6 cm × 6 cm square.	1		6G3a
4	(Diagram with labels: Pentagon, Triangle, Square, Parallelogram, Trapezium)	1	Children need to label: • one of the triangles • the parallelogram • a square • a five-sided shape to demonstrate the pentagon	4G2a
5	Less than 1: $\frac{1}{6} \times 3$, $\frac{1}{3} \times 2$ Equal to 1: $\frac{1}{8} \times 8$ Greater than 1: $\frac{1}{4} \times 5$, $\frac{1}{4} \times 6$, $\frac{1}{9} \times 10$	1		5F5
6	$\frac{3}{10}$ of 20	1		5F6a
7	Sam arrives first at 7.45 p.m. and Sid arrives at 7.50 p.m.	1		5M4
8a	1,250 litres	1		5C6b/6C7c
8b	20 minutes	1		
9a	$x = 25$	1		6A1
9b	$x = 10$	1		
10	7,860	2	**Award 1** mark for correct calculation and incorrect answer e.g. 655 × 12 600 × 12 = 7200 50 × 12 = 600 5 × 12 = 60	6C7a
11	£3,000	2	**Award 1** mark for correct calculations and recognising percentages. £600 = 20% £300 = 10%	6F10

Reasoning Test 9: Mark Scheme

12a	(365 × 3) + (366 × 1) = 1,461 days	3	**Award 1** mark for one correct answer. **Award 2** marks for one correct answer and one correct calculation with incorrect answer.	6C8
12b	$\frac{687}{7}$ = 98 weeks			
Total		18		

Reasoning Test 10: Mark Scheme

Question	Requirement	Mark	Additional guidance	Content domain reference
1	10, 100, 0.1	1		4N3a/4F9
2	16 years	1		4S1
3	160 cm³	1		6M8a/b
4	80°	1		5G2a
5	=	1		5F3
6	(heptagon and hexagon shapes)	1		4F2
7	$\frac{7}{9}$ of £171 is £133 and $\frac{5}{8}$ of £136 is £85. $\frac{7}{9}$ of £171 is worth more than $\frac{5}{8}$ of £136.	1		5G2a
8a	2 minutes	1		5C8a/6C6
8b	11 minutes	1		
9	81 It is the only odd square number between 50 and 100.	2	Award **1 mark** for demonstrating an understanding of square numbers and an example of a square number from 50–100.	5C8a
10a	2 hours 20 minutes	1		6C8/6M9
10b	210 miles	1		
11	650 × 9 = 5,850 5,850 × 0.3 = £1,755	2	Award **1 mark** for demonstrating calculations e.g.: 650 × £9 = £5850 and 5850 ÷ 10 = 585	5F11/6C7a
12a	720°, 900°, 1,080°, 1,260°, 1,440°	3	Award **1 mark** for one correct answer. Award **2 marks** for one correct answer and one correct calculation with incorrect answer.	6G2a/6A3
12b	$(n - 2) \times 180$			
Total		18		

Reasoning Test 11: Mark Scheme

Question	Requirement	Mark	Additional guidance	Content domain reference
1	18, 72, 3, 600	1		3C6
2a	136 bpm	1		4S2
2b	32 bpm			
3	8 cubes	1		5G3b
4	72 cm	1		4M7a
5	Youngest brother £8, middle brother £16 and older brother £16	1		4F10a
6	$9\frac{5}{100}$, $\frac{42}{100}$, 2.5	1		5F8
7	No, 50 + 50 + 50 + 2 is the largest amount.	1		5M9a
8a	24 × 86 = 2,064 books	1		5C7a/5C7b
8b	258 books	1		
9	No, the total is 721, so the average $\frac{721}{5}$ = 144.2	2	Award **1 mark** for adding all heights and dividing by 5 to achieve an incorrect answer.	6S3
10	90 minutes	2	Award **1 mark** for the correct calculation; 72 × 1 = 72 and 72 × 0.25 or 72 ÷ 4 with an incorrect answer.	6F9b
11	£122.50	2	Award **1 mark** for recognising that 98 must be divided by 8.	6F9c/6C8
12a	24 tickets	3	Award **1 mark** for one correct answer. Award **2 marks** for one correct answer and one correct calculation with incorrect answer.	6F9c/6C4
12b	£37.50			
Total		18		

Reasoning Test 12: Mark Scheme

Question	Requirement	Mark	Additional guidance	Content domain reference
1	250	1		4N3b
2a	10%	1		4S2
2b	76			
3	11.6 cm	1		6G5
4	22.5 cm²	1		6M7b
5	140	1		5F12
6	$5\frac{5}{6}$	1		5F2a
7	True. Explanations may vary, for example: adding a negative number to a number will make it smaller. Adding a negative number to a negative number will make it smaller, so the sum will be negative.	1	Children could show this on a number line, starting at a negative number, then adding another negative number by counting on to the left.	5N5
8a	77 drops	1		5C7b/6R1
8b	3 hours 45 minutes	1		
9	870 handshakes	2	Award **1 mark** for recognising the required calculation 29 × 30 and an incorrect answer.	6A2
10	$\frac{5}{12}$	2	Award **1 mark** for demonstrating the correct calculation and working in twelfths. 144 ÷ 12 = 12; 84 ÷ 12 = 7	6F4
11	£4.90	2	Award **1 mark** for demonstrating the calculation of £3.50 ÷ 2.5.	6R1
12a	15.4 °C	3	Award **1 mark** for one correct answer. Award **2 marks** for one correct answer and one correct calculation with incorrect answer.	6C4/6R4
12b	3 times as warm			
Total		18		

Reasoning Test 13: Mark Scheme

Question	Requirement	Mark	Additional guidance	Content domain reference
1	165 168	1		4N4b
2	12:43	1		5S1
3	Any angle that measures more than 90° and less than 180°	1	Examples: 116°, 112°, 168°, 105°, 94°	5G4c
4	(reflected arrow shapes on grid)	1	Shapes must be equal distance from mirror line and the same size.	4G2c
5	$5\frac{1}{8}$	1		5F2a
6	5.93	1		5F10
7	4 pints 4 pints is 2.272 litres	1		5M6
8	315 miles	2	Award **1 mark** for correct calculation of 70 × 4 = 280 and 70 × 0.5 = 35 with an incorrect answer.	6C8/6M9
9a	19, 23	1		6A2
9b	$2n + 3$	1		
10	36.8 m² 2.3 × 2 = 4.6 4.6 × 8 = 36.8	2	Award **1 mark** for demonstrating 2.3 × 2 = 4.6 and 4.6 × 8 = Achieving an incorrect answer.	6M7c
11a	3 : 1	1		6R3
11b	144 slabs (144 : 48)	1		
12a	£7,500	3	Award **1 mark** for one correct answer. Award **2 marks** for one correct answer and one correct calculation with incorrect answer.	6F9c/6C7b
12b	£93.75 ($\frac{4,500}{48} = 93.75$)			
Total		18		

Reasoning Test 14: Mark Scheme

Question	Requirement	Mark	Additional guidance	Content domain reference
1	$\frac{40}{100}$	1		3F1c
2a	Competitor C	1		4S2
2b	67.4 − 65.9 = 1.5 m			
3	93.6 cm	1		5M7a
4	(shape plotted on coordinate grid with vertices approximately at (0,0), (3,0), (1,3), (3,6), (0,6))	1		4P3a
5	$\frac{3}{4}, \frac{3}{10}, \frac{1}{4}, \frac{1}{8}$ or fraction equivalents	1		5F6b
6	$\frac{3}{4}, (\frac{6}{4} - \frac{3}{4})$	1		5F4
7	6 days. 6 days is 6 × 24 hours = 144 hours 144 > 140	1		5M4
8a	Yes. The price for 6 people would be 30 + 8.50 + 8.50 = £47.00.	1		5C8a
8b	£4	1		
9	60.3	2	Award **1 mark** to demonstrate understanding of missing values with one correct calculation and one incorrect e.g. 3 × 7.5 = 22.5 22.5 = 7 = 29.5 4 × 9.2 = 36.8 − 6	6A1
10a	742.5 g	1		6R1
10b	1:11	1		
11a	16 balls	1		6C7c/6R2
11b	£58.50	1		
12a	To buy the phone outright because the monthly contract is £360	3	Award **1 mark** for one correct answer. Award **2 marks** for one correct answer and one correct calculation with incorrect answer.	6C8
12b	Talk costs 10p per minute			
Total		18		

Reasoning Test 15: Mark Scheme

Question	Requirement	Mark	Additional guidance	Content domain reference
1	450, 99, 117	1		4N1
2	£4.83	1		4S2
3	21 cm²	1		5M7b
4	10.8 cm	1		6G5/5C4
5	$7\frac{5}{7}$	1		5F2a
6	Two of $\frac{16}{20}$, $\frac{8}{10}$, $\frac{40}{50}$, $\frac{80}{100}$ Or any other two fractions equivalent to $\frac{4}{5}$.	1		5F2b
7	5 hours 45 minutes	1		5M4
8	5 lb 5 oz	2	Award **2 marks** for correct answer. Award **1 mark** for correct working and an incorrect answer.	5M6
9	11, 12 and 13 The number will be above 30 because every -teen number is made of a ten and ones. The number cannot exceed 54 because 17, 18 and 19 are the highest possible numbers. Therefore it can only be 36 or 49.	2	Award **2 marks** for correct answer. Award **1 mark** for evidence of calculations to identity the possibilities 36 and/or 49. Children should demonstrate a strategy to calculate the possibilities.	5C5d
10a	4,800 m, 420	1	Award **2 marks** for two correct answers.	6C8
10b	No: 4 × 12 = 48 + (30 seconds × 12 = 6 minutes) = 54 minutes which is more than 45 minutes.	1		
11a	31 m	1	Award **2 marks** for two correct answers	6M7c/6C7a
11b	480.5 m²	1		
12a	32 × 8 = 256 apples	3	Award **2 marks** for 256 apples and 2.25 × 100 = 225. Award **3 marks** when 225 is divided by 2 to get the correct answer.	6R1/6C7a
12b	2.25 × 100 = 225.00 225 ÷ 2 = £112.50			
Total		18		

Reasoning Test 16: Mark Scheme

Question	Requirement	Mark	Additional guidance	Content domain reference
1	The second jug in the top row and the bottom two jugs should be circled.	1		3M1c
2	23°C	1		3M2a/3S2
3	Lines must be 1 cm apart all the way along; line A must be within 1 mm of 7.5 cm and line B must be within 1 mm of 4.5 cm.	1		3G2
4	A four-sided shape with two angles measuring less than 90° and two measuring more than 90° but less than 180°	1		4G4
5	$\frac{3}{4}$	1		5F4
6	$\frac{3}{8}$	1		5F2b
7	False, the mean average is 51	1		5C5d/6S3
8a	$\frac{1}{4}$ are lettuces	1		5C5a/5F4
8b	18	1		
9	2, 3, 4, 9 2 × 4 = 8 3 × 9 = 27 2 × 3 = 6 4 × 9 = 36	2		5C6a
10	3,550 × 5 = 17,750 3,550 × 0.5 = 1,775 1,775 + 17,750 = £19,525	2	Award **1 mark** for demonstrating the correct calculations 3550 × 5 = 17750 3550 × 0.5 = 1775 With an incorrect total.	6C7a
11a	3 cm	1		6R3/6M7b
11b	5.25 × 4 = 21 cm² or 10.5 × 2 = 21 cm²	1		
12a	40 go on the coach and 10 go with parents.	3	Award **1 mark** for one correct answer. Award **2 marks** for one correct answer and one correct calculation with incorrect answer.	6R1/6F6
12b	$\frac{3}{10}$ = 15 pupils 15 × 6 = 90 90 − 0.15 = £89.85			
Total		18		

Reasoning Test 17: Mark Scheme

Question	Requirement	Mark	Additional guidance	Content domain reference
1	80%	1		4F10
2	UK: 66 million South Africa: 56 million Spain: 46 million Canada: 36 million Australia: 24 million Bar chart representing these rounded numbers accurately.	1		5N4/5S1
3	4 hours 49 minutes	1		4M4c
4	120°	1		6G4a
5	Number of quarters in 1.5 = 6	1		5F2a
6	$0.8 = \frac{4}{5}$ $0.375 = \frac{3}{8}$ $0.35 = \frac{35}{100}$	1		5F6b
7	840 seconds (14 minutes) > 13 minutes (780 seconds)	1		5M4
8a	$35	1		5M9a
8b	£15	1		
9	$\frac{24}{3} = 8$ 24 + 8 = 32 children	2	Award **1 mark** for the calculation 24 ÷ 8 = 3 demonstrating an understanding that boys = $\frac{3}{4}$.	6F4
10a	1.4 × 3 = 4.2 kg 4.2 kg × 26 = 109.2 kg	1		5M5/6M9
10b	1.15 × 3 = £3.45 3.45 × 3 = £10.35	1		
11a	3,200 × 0.8 = £2,560	1		6F11/6F9
11b	3,200 × 1.15 = £3,680	1		
12a	$\frac{930}{8} = 116.25$ 116.25 × 5 = 581.25 miles	3	Award **1 mark** for one correct answer. Award **2 marks** for one correct answer and one correct calculation with incorrect answer.	6M6/5C7b
12b	103 km/h			
Total		18		

Reasoning Test 18: Mark Scheme

Question	Requirement	Mark	Additional guidance	Content domain reference
1	150, 175, 200	1		4N1
2	$1.73 (Aus$)	1		4S2
3	3 + 3 + 7 + 3 + 10 + 10 + 3 + 7 + 3 + 3 = 52 m	1		4M7a
4	(clock showing approximately 7:33)	1		4M4b
5	35 10.5 ÷ 3 = 3.5, 3.5 × 10 = 35	1		5F5
6	$\frac{20}{30}, \frac{12}{30}, \frac{10}{30}, \frac{4}{30}, \frac{27}{30}$	1		5F2b
7	10 cm I know that 25 cm² = 25% of the area of square B. 25 × 4 = 100 and 100 = 10 × 10 So the side length of square B is 10 cm.	1		5M9a
8	6 × 6 × 6 = 216 cm³ 3 × 3 = 9 $\frac{216}{9}$ = 24 cm The water will reach a height of 24 cm.	2	Award **1 mark** for demonstrating 6 × 6 × 6 = 216 and 3 × 3 = 9 with further incorrect calculations.	6M8a
9	72 has factors 36, 24, 18, 12, 9, 8, 6, 4, 3, 2.	2	Award **1 mark** for demonstrating 8 factors of 72.	5C5a
10	45 cm and 75 cm	2		6R1
11	620 ÷ 1.6 = 387.5 miles	2	Award **1 mark** for demonstrating an understanding of how to calculate 620 ÷ 1.6 for example how many 1600s in 620,000.	6M6
12a	$\frac{425}{8.5}$ = 50 mph	3	Award **1 mark** for one correct answer. Award **2 marks** for one correct answer and one correct calculation with incorrect answer.	6C8
12b	$\frac{425}{5}$ = 85 85 × 1.34 = £113.90			
Total		18		

Reasoning Test 19: Mark Scheme

Question	Requirement	Mark	Additional guidance	Content domain reference
1	$\frac{1}{3}, \frac{1}{2}, \frac{2}{3}, \frac{5}{6}, \frac{6}{6}$	1		5F3
2	375 cars were sold on average per month from January to April.	1		4S2/6S3
3	Hexagonal prism	1		5G3b
4	The pentagon should be drawn in this orientation.	1		5P2
5	$\frac{5}{9}$	1		5F5/5F4
6	$32\frac{3}{4}$	1		5F2a
7	75 $\frac{2,250}{30} = 75$ $75 \times 30 = 2,250$	1		5C8a
8a	$9 \times 16 = £1.44$ so 9p per slice	1		5F10/5C8a
8b	$12.5 \times 16 = £2$	1		
9a	$86.4 + 2.8 = 89.2$	1		6A1/6C8
9b	$5 - 0.27 = 4.73$	1		
10a	£4,725 10% = 350 $350 \times 3 = 1,050$ 5% = 350 ÷ 2 = 175 $3,500 + 1,050 + 175 = £4,725$ $3,500 \times 1.35 = £4,725$	1		6F9a/6F11
10b	20%	1		
11a	36 cm	1		6C7a/6M7a
11b	$48 \times 36 = 1,728$ cm²	1		
12a	$23 \times 349 = £8,027$	3	Award **1 Mark** for one correct answer. Award **2 Marks** for one correct answer and one correct calculation with incorrect answer. $\frac{389}{10} = 38.9$ $38.9 \times 4 = 155.60$ $155.60 + 19.45 = 175.05$ $175.05 \times 10 = £1,750.50$	6M9/6F11/6C8
12b	45% of £3,890 = £1,750.50			
Total		18		

Reasoning Test 20: Mark Scheme

Question	Requirement	Mark	Additional guidance	Content domain reference
1	13 mm, 25 mm, 12 cm, 1.2 m, 1.8 m	1		3M1a
2	8.3 m 8.59 + 8.16 + 8.15 = 24.9 $\frac{24.9}{3} = 8.3$	1		4S1/6S3
3	24.5 cm^2 3.5 × 7 = 24.5	1		5C6a/6M7b
4	216 cm^3 6 × 6 × 6 = 216	1		6M8b
5	0.25	1		5F6a
6	a) False b) True c) False	1		5F3
7	Heidi is correct. A square has the same properties as a rectangle: • 4 right angles • 4 sides • 2 sets of parallel sides, but also: • Opposite sides of equal length	1		5G2a
8a	187.5 ml	1		5F10/6R1
8b	750 ml	1		
9	17 apples	2	Award **1 mark** for evidence of working backwards.	6C8
10	190 cm^2 3 × 5 = 15 10 × 5 = 50 10 × 3 = 30 30 + 50 + 15 = 95 95 × 2 = 190	2	Award **1 mark** for demonstrating the correct calculation e.g. 3 × 5 = 15 10 × 5 = 50 10 × 3 = 30 30 + 50 + 15 = 95 95 × 2 = 190	5M7b/5C6a
11a	40 cubes	1		6G2a/6R1
11b	140 cubes	1		
12a	£12,750 255,000 ÷ 100 = 2,550 (1%) 2,550 × 5 = 12,750	3		6C8
12b	£280,500 25 × 12 = 300 300 × 935 = 280,500		100 × 935 = 93,500 100 × 935 = 93,500 100 × 935 = 93,500 3 × 93,500 = 280,500. Award **2 marks** for two correct strategies but incorrect answer. Award **1 mark** for one correct strategy and working and incorrect answer.	
Total		18		

Reasoning Test 21: Mark Scheme

Question	Requirement	Mark	Additional guidance	Content domain reference
1	$0.7 \rightarrow \frac{70}{100}$ $0.007 \rightarrow \frac{7}{1,000}$ $7 \rightarrow \frac{70}{10}$ $0.07 \rightarrow \frac{7}{100}$	1		4F6b
2	$\frac{12}{30}, \frac{4}{10}$ or $\frac{2}{5}$	1		4S2
3	729 cm³	1		4M9
4		1	Any three sides added to make an octagon.	5G2b
5	42 L	1		5F4
6	Any fraction equivalent to $\frac{3}{4}$, for example: $\frac{9}{12}, \frac{15}{20}, \frac{30}{40}, \frac{75}{100}$	1	Divide the fraction with a calculator to check, for example, 9 ÷ 12 = 0.75 If the answer = 0.75 the fraction is correct.	5F2b
7	No. The washing will not finish until 6.17 p.m. 3.35 count on 2 hours = 5.35 5.35 add on 30 = 6.05 6.05 + 12 = 6.17 p.m.	1		5M4
8	8 minutes	2	Award **1 Mark** for 1 hour 44 minutes = 104 minutes, $\frac{104}{13}$ = incorrect answer	6M9
9	48 × 1, P = 98 cm 24 × 2, P = 52 cm 16 × 3, P = 38 cm 12 × 4, P = 32 cm 8 × 6, P = 28 cm	2	Award **1 mark** for finding $\frac{4}{5}$ possibilities.	6M7a
10a	10:50 a.m.	1		6M9/6R1
10b	1 hour 25 minutes	1		
11	15 blue, 20 red, 25 green	2		6R1
12a	£975 per week	3	37.5 × 26 = 975 Award **2 marks** for this stage.	6M9/6C8
12b	13 weeks 390 × 13 = 5,070 and he needs 5,000		Award **1 mark** for part b) if correct with explanation: Yes because 390 × 13 = 5,070 and he needs 5,000.	
Total		18		

Reasoning Test 22: Mark Scheme

Question	Requirement	Mark	Additional guidance	Content domain reference
1	6, 10, 15, 21	1	Triangular numbers – 3, 6, 10, 15, 21	3C4
2a	Liverpool	1		3S2
2b	24 points			
3	3 lines of symmetry correctly drawn	1		4G2b
4	Yes	1		4G2a
5	56	1	$\frac{1}{5}$ of 70 = 70 ÷ 5 = 14 $14 \times 4 = 56$	5F5
6	$\frac{3}{9}, \frac{3}{4}, \frac{89}{100}$	1		5F2a
7	Different angles. One shape has 2 sets of parallel sides, the other only has 1 set.	1		5G2b
8	144 slabs	2	$4.5 \times 8 = 36 \text{ m}^2$ 4 slabs = 1 m² $4 \times 36 = 144$ slabs	6M7b
9	68	2	24 + 68 = 92 $\frac{92}{2} = 46$	6C4
10a	3.48 a.m.	1	3.48 a.m. (must say a.m. or 03:48)	6M9
10b	3,250 miles	1		
11	3 litres. She needs 4 × 568 ml = 2,272 ml = 2.272 litres. So she will need to buy 3 one-litre boxes	2		6M9
12a	£5.82	3		6F10
12b	£1,613.60		£25.99 – £5.82 = £20.17 20.17 × 80 = £1,613.60	
Total		18		

Reasoning Test 23: Mark Scheme

Question	Requirement	Mark	Additional guidance	Content domain reference
1	96, 48 and 24	1		3C6
2a	40	1		3S2
2b	5		6 − 1 = 5	
3	24 cm²	1	3 × 4 = 12 cm² 12 cm² × 2 = 24 cm²	6M7b
4	Any angle measuring more than 180° and less than 360°. Angle must be labelled.	1		5G4a
5	$\frac{11}{15}$	1	$\frac{15}{45} + \frac{18}{45} = \frac{33}{45}$ $\frac{5}{15} + \frac{6}{15} = \frac{11}{15}$	6F4
6	348	1		6F11
7	Because three odd numbers will always add to an odd number (for example, 1 + 1 + 1 = 3).	1		6C6/6C3
8	441 cm²	2	294 × 1.5 = 294 ÷ 2 = 147 147 + 294 = 441 Award **1 mark** for a correct calculation with an incorrect final answer.	6R3
9	10, 17, 24	2		6C4
10	180 children	2	$\frac{420}{7} = 60$ 60 × 3 = 180 180 children	6F9c
11	A cube A = 125 cm³ and cuboid B = 120 cm³ Therefore A has the greater volume.	2		6M8a
12a	Yes	3	13.33 minutes later = 13 minutes 20 seconds 6.30 + 13.20 = 6.43 20 seconds (yes)	6M9/6C8
12b	8.64p		0.108 × 10 = 1.08 1.08 × 8 = 8.64p	
Total		18		

Reasoning Test 24: Mark Scheme

Question	Requirement	Mark	Additional guidance	Content domain reference
1	Shape is in new position, in 12 o'clock position above A, on its side.	1		5P2
2a	45	1		5S1
2b	$\frac{1}{4}$ or $\frac{30}{120}$			
3	A = Circumference B = Diameter C = Radius	1		6G5
4	Rhombus	1		5G2b
5	$\frac{5}{6}$	1		6F2
6	2,500	1		5F6b
7	13	1		5C6a
8	£2,040	2	12 × 4 = 48 12 × 6 = 72 48 + 72 = 120 120 × 10 = 1,200 120 × 7 = 700 + 140 = 840 1,200 + 840 = 2,040	5C7a
9	Four polygons with more than four sides named, for example: pentagon, hexagon heptagon, octagon, etc. Any reasonable explanation, such as that given in Additional guidance column.	2	A polygon is a closed shape with 3 or more sides, so it must be that. There are 4 visible straight edges which means it must be at least a quadrilateral. However two sides are parallel, which means there has to be another side to join them making it a polygon with at least 5 sides.	6G2a
10	£41.25	2	11.75 × 11 = £129.25 15.50 × 11 = 170.50 170.50 − 129.25 = £41.25	6M9
11	660 rungs	2	12 × 45 = 540 8 × 15 = 120 660	6C7a
12a	Cat = 25 + (5 × 4) = 45	3	Award **2 marks** for a correct answer to part a) and 1 further mark if calculations for horse are accurate with incorrect answer.	6R1
12b	66.5 − 6.5 years (3 human years) = 60 horse years 60 years ÷ 5 (human years) = 12 12 human years + 3 human years = 15 human years.			
Total		18		

Reasoning Test 25: Mark Scheme

Question	Requirement	Mark	Additional guidance	Content domain reference
1	16	1		4F1
2	140 minutes	1		4M5
3	48 cm^2	1	16 cm^2 + 32 cm^2 = 48 cm^2	6M7b
4	50°	1		6G4a
5	$\frac{1}{10}$	1		5F4
6	75	1		5F10
7	No 30 × 3.28 = 98.4 ft The school field is not 100 ft wide therefore not big enough for the helicopter to land.	1	40 × 3.28 = 131.2 ft	5M6
8	1 minute 54 and 26 hundredths seconds, or 1 minute 54.26 seconds	2	57.13 + 57.13 = 114.26 = 1 minute 54 and 26 hundredths seconds (1:54:26)	5M4
9	275	2	(700 − 150) ÷ 2 = 275 275 + 275 + 150 = 700 275 + 425 = 700	6C4
10a	18,000 cm^2	1	18,000 cm^2	6M7b/6C8
10b	£252 profit	1	360 × 0.7 = £252	
11	2,555 pages	2		5C7a
12a	2,640 logs	3	48 × 55 = 2,640	6C7a
12b	£90		7.5 × 12 = £90	
Total		18		

Reasoning Test 26: Mark Scheme

Question	Requirement	Mark	Additional guidance	Content domain reference
1	Divisible by 4: 88, 72, 16 Divisible by 5: 15, 85 Divisible by both: 80, 100, 20	1		3C4
2	10:15	1		3S2
3	Check children's drawing.	1	Cuboid with Length 5 cm, Width 3 cm, Height 2 cm	6G3a
4	Decagon	1	Accept also five-pointed star, pentagram or star pentagram.	5G2b
5	$31\frac{2}{3}$	1		5F5
6	60%	1		5F5
7	Elijah is right because 112 × 7 = 784 and 876 ÷ 8 = 109.5. They are not equal.	1		6C8
8a	approximately 4,500 feet	1	1,350 × 3 = 4,050 $\frac{1}{3}$ of 1,350 = 450 4,050 + 450 = 4,500	6M5
8b	$4\frac{1}{2}$ hours	1		
9	52.5 minutes. There will be five rounds with a break between each (four breaks).	2	8.5 × 5 = 42.5 2.5 × 4 = 10 42.5 + 10 = 52.5	5M4
10a	18 red, 36 white, 45 blue	1	18, 36, 45 12 × 20 = 240 9 × 50 = 450	6R1/5M9a
10b	£6.90	1	2.40 + 4.50 = 6.90	
11	333 days	2	333.33… days rounded to nearest whole number 333 days.	5C7b
12a	150.5 cm	3	150.5 cm	6R3/6M9
12b	127 cm		4 ft 2 inches = 50 inches 50 × 2.54 = 127 cm	
Total		18		

Reasoning Test 27: Mark Scheme

Question	Requirement	Mark	Additional guidance	Content domain reference
1	75 seconds	1		4M4c
2a	£225,000	1		3S2
2b	£49,500	1		
3	An angle between 100° and 130° and another angle between 50° and 80°.	1	Must be drawn with a ruler.	4G4/5G4c
4		1		4G3a
5	A circled.	1	$\frac{11}{11} + \frac{15}{20} = 1\frac{3}{4}$, $\frac{4}{4} + \frac{3}{4} = 1\frac{3}{4}$ Equation B balances.	5F4
6	8 parts shaded	1		5F2b
7	It has 12 sides and 12 angles	1		6G2a
8	Alex Alex earns 2,200 × 12 = £26,400 Brad earns 500 × 52 = £26,000	2		6C7a
9	HTT, HTH, HHT, HHH, TTT, THT, THH, TTH	2	Award 1 mark for 4 correct (excluding HHH)	6S1
10a	4,800 seconds	1	80 × 60 = 4,800	6M5
10b	40 hours	1	$\frac{2,400}{60}$ = 40 hours Or 30 × 1hr 20 minutes	
11a	52 cats are tabby	1	$\frac{130}{10}$ = 13 13 × 4 = 52	6R2
11b	$\frac{1}{10}$ or 10%	1	40 + 30 + 20 = 90% so 10% are other colours.	
12a	13, 16, 19, 22, 31, 301	3		6A3
12b	$3n + 1$ or 3 × number of squares plus one			
Total		18		

Reasoning Test 28: Mark Scheme

Question	Requirement	Mark	Additional guidance	Content domain reference
1	245	1		4M4c
2	125 cm	1	122 + 118 + 125 + 140 + 120 = 625 $\frac{625}{5}$ = 125 cm	4S2/6S3
3	Any polygon with 3–8 sides, in which one angle measures 120°/121°/119°.	1	Allow 1 degree of inaccuracy.	5G4c
4	70 cm^2	1	12 × 5 = 60 cm^2 2.5 × 4 = 10 cm^2 70 cm^2	5M7b/6M7b
5	Any one of $\frac{12}{36} > \frac{3}{12}$ $0.6 > \frac{3}{12}$ $0.6 > \frac{12}{36}$	1		6F11
6	$\frac{36}{54}$	1		5F2b
7	Leah 10 and Poppy 15	1		6C8
8	8	2	8 × 1.25 and 4 × 0.75	6M9
9	128 m^2	2		6M7c
10a	17 °C	1		6M9
10b	12 degrees	1		
11a	290 boys	1	725 × $\frac{2}{5}$ = 290	6R4
11b	240 girls	1	400 × $\frac{3}{5}$ = 240	
12a	Any combination of 2 starters, 2 mains and 2 desserts.	3		6C4
12b	£55.93			
Total		18		

Reasoning Test 29: Mark Scheme

Question	Requirement	Mark	Additional guidance	Content domain reference
1	65 kg → 143 lb 10 kg → 22 lb 1 kg → 2.2 lb 15 kg → 33 lb	1		4M5
2	£46.80	1	172.40 − 125.60 = £46.80	4S2
3	Any one rectangle from 1 × 36, 2 × 18, 4 × 9, 6 × 6	1	Any of these options drawn according to grid scale.	4M7b
4	Trapezium	1		4P3b
5	78	1	42 + 36 = 78	5F11
6	$1\frac{7}{15}$	1	$5 \times 3 = 15$ $4 \times 3 = 12$ $2 \times 5 = 10$ $\frac{12}{15} + \frac{10}{15} = \frac{22}{15}$	5F2a
7	No	1	No, both jumpers cost more than £20 Jumper A = £21 Jumper B = £20.30	5F11
8a	20 doses	1		6C8
8b	26 doses	1		
9	12 of cube A	2	$5 \times 1.8 = 9$ cm $12 \times 2.5 = 30$ cm	6CM9
10	MCCCXVI	2	$28 \times 47 = 1{,}316$	5N3b/6C4
11a	36 cm	1		6M7a/6R3
11b	729 cm^2	1		
12a	Smoothie = £2.25 Flapjack = £1.25	3	20% of shop price is 30p per flapjack.	6C8
12b	£210		700 × 30p = £210	
Total		18		

Reasoning Test 30: Mark Scheme

Question	Requirement	Mark	Additional guidance	Content domain reference
1	34	1		4N3b
2	1,625 books	1		4S2
3	49.5 g	1	$9 \times 5 = 45$ $9 \times 0.5 = 4.5$	5M9c
4	The shape drawn in the correct position (below A) and orientation (as shown).	1		5P2
5	$\frac{1}{12}$	1	$\frac{1}{3}$ = 4 parts $\frac{3}{3}$ = 12 parts Therefore $\frac{1}{3}$ shared between 4 is $\frac{1}{12}$ Shape should be divided into 12 parts	6F5b
6	$\frac{7}{24}$	1	$\frac{5}{8} = \frac{15}{24}$ $\frac{1}{3} = \frac{8}{24}$ $\frac{15}{24} - \frac{8}{24} = \frac{7}{24}$	6F2
7	Tom	1	$6 \times 1,600 = 9,600$ $9,600 < 10,500$ Therefore Tom went further.	6M6
8a	200 g	1	$40 \times 5 = 200$ g	6R1
8b	21 cups	1	$\frac{10.50}{0.50} = 21$ cups	
9	$x = 12$	2	$x = 12$ $\frac{84}{2} = 42$ $42 - 18 = 24$ $\frac{24}{2} = 12$	6A2
10a	Idris 20 seconds per length Kala 15 seconds per length	1		6S3
10b	Yes	1		
11	No, they need another 12 volunteers.	2	$\frac{480}{10} = 48$ $48 \times 7 = 280 + 56 = 336$	6F11

Reasoning Test 30: Mark Scheme

12a	No, she will arrive at 9.25 a.m.	3		6M9
12b	£0.90 or 90p		8.46 × 15 = £126.90	
Total		18		

Name _____   Class _____

Year 6/P7 Weekly Reasoning Tests Record Sheet

Tests	Mark	Total marks	Key skills to target
Test 1			
Test 2			
Test 3			
Test 4			
Test 5			
Test 6			
Test 7			
Test 8			
Test 9			
Test 10			
Test 11			
Test 12			
Test 13			
Test 14			

Name _____   Class _____

Test 15			
Test 16			
Test 17			
Test 18			
Test 19			
Test 20			
Test 21			
Test 22			
Test 23			
Test 24			
Test 25			
Test 26			
Test 27			
Test 28			
Test 29			
Test 30			

© HarperCollins*Publishers* Ltd 2019

www.ingramcontent.com/pod-product-compliance
Lightning Source LLC
Chambersburg PA
CBHW081436300426
44108CB00016BA/2378